25–
X

MOLECULES, DYNAMICS, AND LIFE

NONEQUILIBRIUM PROBLEMS IN THE PHYSICAL SCIENCES AND BIOLOGY

Editors: I. Prigogine and G. Nicolis

Université Libre de Bruxelles
Brussels, Belgium

MOLECULES, DYNAMICS, AND LIFE

An Introduction to Self-Organization of Matter

A. BABLOYANTZ

University of Brussels
Brussels, Belgium

A WILEY-INTERSCIENCE PUBLICATION

JOHN WILEY & SONS

New York · Chichester · Brisbane · Toronto · Singapore

To Eric and Michel

Library of Congress Cataloging in Publication Data:

Babloyantz, A. (Agnessa)
    Molecules, dynamics, and life.

    (Nonequilibrium problems in the physical sciences and biology,
    ISSN 0275-9292; v. 4)
    "A Wiley-Interscience publication."
    Bibliography: p.
    Includes index.
    1. Molecular biology.   2. Thermodynamics.
3. Biophysics.   I. Title.   II. Series.

QH506.B33   1986   574.8'8   85-26413
ISBN   0-471-82380-5

Printed in the United States of America

10 9 8 7 6 5 4 3 2 1

FOREWORD

It is now nearly 40 years since I published a monograph, *Etude thermodynamique des phénomènes irréversibles* (See Chapter 4), devoted specifically to thermodynamics of nonequilibrium and self-organization.

This was a slim book. Very little was known outside equilibrium state. Today, the intellectual landscape has undergone drastic changes. A fundamental reconceptualization of science is going on. We have long known that we are living in a pluralistic world in which we find deterministic as well as stochastic phenomena, reversible as well as irreversible. We observe deterministic phenomena such as the frictionless pendulum or the trajectory of the moon around the earth; moreover, we know that the frictionless pendulum is also reversible, as future and past play the same role in the equations describing the motion of the pendulum. But other processes, such as diffusion or chemical reactions, are irreversible. Here, there is a privileged direction of time: the system becomes uniform after some time. We must also acknowledge the existence of stochastic processes if we want to avoid the paradox of referring to the variety of natural phenomena as a blueprint designed at the very moment of the Big Bang.

What has changed since the beginning of this century is our evaluation of the relative importance of these types of phenomena.

The artificial may be deterministic and reversible. The natural contains essential elements of randomness and irreversibility.

At the start of this century, continuing the tradition of the classical approach, physicists were almost unanimous in admitting that the fundamental laws of the universe were deterministic and reversible. Processes which did not fit this scheme were supposed to be exceptions, mere artifacts of some apparent complexity, which itself had to be accounted for by invoking our ignorance, or lack of control over the variables involved. Now that we are at the end of this century, we are more and more inclined to think that the fundamental laws of nature are irreversible and stochastic; that deterministic and reversible laws are applicable only in limited situations.

This change in our outlook on the world has been the subject of numerous monographs and review articles. However, the emphasis is generally put on the formal part of the theory, closely related to the advances in our understanding of unstable and nonlinear dynamic systems. This fact clearly makes these books difficult to read for most members of the scientific community.

I am therefore very happy to present this book by Agnessa Babloyantz, in which the emphasis is more on the physical content than on the formal aspects of this new outlook. I believe we have to thank Babloyantz for having written a highly readable text, accessible to everyone who is interested in the natural sciences.

As a result of the new ideas developed in the field of nonequilibrium systems, the gap between subjects which were traditionally considered to be "simple" and the ones—such as biology or human sciences—which were thought of as complex is becoming narrower. This leads to the possibility of the transfer of knowledge from one field to another. For this very reason, the present book should be of interest to researchers in fields such as physics, chemistry, biology, and social sciences.

Agnessa Babloyantz is highly qualified to write this book: she has published valuable work devoted to nonequilibrium systems in biology, including significant contributions to morphogenesis, and more recently to the dynamical approach of neural activity, including a study of the fractal geometry of attractors in brain dynamics.

Her recent work has shown that the electrical activity of the brain in deep sleep as monitored by electroencephalogram (EEG) may be modeled by a "fractal attractor." This is remarkable as it shows that the brain acts as a system possessing intrinsic complexity and unpredictability. It is this instability which likely permits the amplification of inputs due to sensory impressions in the waking state.

In spite of the spectacular progress made these last years in nonequilibrium physics and mathematics, in spite of the innumerable conference proceedings and papers which are being published, we are only at the beginning of a new dialog with nature. It may be hoped that this book by Agnessa Babloyantz will be a reliable guide to help students and research workers find their way in this new area.

I. Prigogine

PREFACE

This book sets out to tell the story of how inert matter can acquire self-organizing and other properties usually ascribed to life. There are many excellent popular books in the various fields of science which present to the reader all the magic of knowledge in the manner of an exciting thriller. The reader is not invited behind the scenes but is only shown the finished results, which might have taken a few days or a few centuries to realize.

The next stage in scientific writing is the elementary textbook. In such books the author chooses a well-defined subject and guides the reader through the topic. Step by step the reader discovers the methods, the logic, the rigor, and also the pitfalls and limitations of a particular scientific endeavor. Such textbooks usually require some initial background, a substantial amount of time, and a great deal of motivation.

In this book I try a middle of the road approach, halfway between a popular work and a textbook. As in popular works, the reader is not required to have a solid background in any of the subjects which are discussed. Every scientific concept is explained in common words. The textbook model is reflected in the systematic approach to each subject. Since the book is conceived as a self-sufficient entity, a special effort has been made to incorporate all the necessary information required for an understanding of its main concepts: the self-organization of matter and its relation to living organisms.

The reader is invited behind the scenes, introduced to the kind of questions asked in a scientific endeavor, and shown how answers are found to those questions. Using logic, a scientist accumulates experimental results, generalizes findings into laws, and predicts the future from the present. Whenever possible, I invite the reader to participate in this scientific process. As in a textbook the reader is shown how to solve simple problems. For more complex situations, only the final results are given and are discussed in plain language. Historical facts, dates, and the names of well-known scientists are incorporated into the main body of the narrative, to give some idea of the historical development of the main concepts.

This book is intended for students interested in natural science. It may also interest physicists, chemists, biologists, engineers, and those who desire to learn without too much effort what has happened in their field since they left college. The reader must have some elementary notions of algebra, and must be familiar with the concepts of differentials, integrals, and differential equations.

This book is composed of three parts, which are connected by the logic of the concept of self-organization.

The first chapter of Part I begins with a few words on the history of the evolution of our ideas concerning matter. It continues with a summary of our present concept of the atom, its internal structure, and the reasons for the association of atoms into molecular assemblies. The details of molecular encounters in the course of a reaction and the amount of change of substances that appear or disappear per second are crucial for the understanding of the self-organizational processes presented in the remaining chapters of the book. Therefore, a relatively large amount of space is devoted to the field of chemical kinetics, which is treated in some detail in Chapter 2. I show how the rate of a chemical reaction may be evaluated. Various elementary reactions are defined; and I indicate how they can be described and studied by mathematical tools.

The concept of entropy and the law which governs its evolution toward a maximum value is the second law of a branch of macroscopic physics called *thermodynamics*. The development of thermodynamics was triggered by the invention of heat engines. Later, it appeared that thermodynamic methods could be used with great profit for the study of bulk matter, particularly chemical and biochemical processes. In some of their aspects a human body and a machine obey the same laws. Thermodynamic methods are one of the most natural tools for the investigation of macroscopic systems since they are based on the observable and tangible properties of bulk matter. To single out clearly the recent important developments in the field of thermodynamics, the subject is divided quite unusually into four separate parts, which are presented in Chapters 3 to 6 of Part I.

In Chapter 3 I define the important concepts of equilibrium thermodynamics and introduce its four laws. It is shown that the thermodynamics of quiescent systems can be extended to the domain of chemical and biochemical reactions, and therefore to the study of living organisms.

In Chapters 4, 5, and 6 the thermodynamic formalism is extended to irreversible processes. In Chapters 3 and 4 thermodynamic methods are used for the study of the measurable properties of bulk matter. For example, one may evaluate the quantity of energy (calories) contained in a sugar cube or measure the speed with which matter crosses cell walls.

Thermodynamic methods also have predictive value. They indicate the direction of evolution of processes. For example, the evolutionary laws of thermodynamics have convinced scientists once and for all that one cannot design airplanes that can fly with the energy of the ambient air as their unique source of power, or run a boat with power extracted solely from ocean waters.

Thermodynamics may be used in a third way. With the help of its methods one can test the stability of a given state of matter under given external conditions. Thermodynamic stability criteria tell us what may happen if small disturbances are introduced either by internal fluctuations or by external perturbations in the state of matter.

The recent advances in the theory of thermodynamics of irreversible processes achieved by P. Glansdorff and I. Prigogine have been primarily in the area of the evolution and stability of irreversible processes. In particular, the development of stability criteria for open systems gives the conditions under which the matter can self-organize spontaneously into temporal and spatial structures.

These chapters devoted to thermodynamics stand as a separate entity and may be read independently from the remainder of the book. They will be of special value to engineers and physicists interested in new developments in this field.

Once the self-organization of matter became compatible with the macroscopic laws of physics, researchers were sufficiently motivated to try alternative, nonthermodynamic methods for the study of the behavior of such systems. Kinetic methods for the investigation of dissipative structures are developed in Part II. The reader not particularly keen on thermodynamic concepts and only interested in acquiring a working knowledge of self-organizing systems may omit the discussion of thermodynamic theory and proceed to Part II.

In Part II, Chapter 7 starts with a survey of self-organizational phenomena in various areas of the natural sciences. It shows how extremely diverse and seemingly unrelated phenomena turned out to be governed by a few common principles. Open, nonlinear, and cooperative systems may give rise to self-organized dissipative structures.

The Belousov–Zhabotinski reaction, taken as a prototype of a dissipative structure, is studied in Chapter 8. The origin of oscillations, wave propagation and stationary structures, chaotic behavior, and many other strange properties of this reaction are shown. These unexpected properties open new fields in chemical sciences, and make it possible for us to speak of a "new chemistry."

In Chapter 9 we describe the self-organizational phenomena in terms of a theoretical approach. Methods are given whereby one can easily detect the presence of dissipative structures in chemical reactions. Model reactions are constructed and their time evolution is followed with the help of differential equations and their solutions.

Today the role of self-organization and dissipative structures in the study of living matter is widely recognized. These concepts have entered the fields of physics, chemistry, biology, population dynamics, and even the social sciences, and have created a multidisciplinary field of research.

To give a comprehensive idea of all the important advances accomplished in the field of self-organized systems would require several volumes. The choice of illustrative examples has not been easy. However, to justify the title of the book, examples had to be taken from the biological sciences. The following criteria were used for selection. Each example illustrates one particular aspect of a living organism that can be traced to one of the self-organizational properties of chemical dissipative structures. All the reported models are studied with the mathematical tools developed in Part II. Finally, many of the examples have been developed by the Brussels school where the new advances in thermodynamics and the ensuing concepts of dissipative structures were developed.

All the examples treated in Part III use mathematical language. The reader is not required to possess a background in biochemistry; all the concepts needed to understand a particular example are furnished in the relevant sections.

Part III begins with a short summary of the evolution of our ideas concerning living organisms, and continues with an account of the structural aspect of important present-day biological macromolecules, indicating how these evolved from simple molecules. With M. Eigen we follow the Darwinian-type evolution of these molecules into *hypercycles*. These complex and structured entities, following a compartmentalization process, probably gave rise to the first primitive living protocells.

The example treated in Chapter 11 shows how glucose consumption by unicellular yeast cells follows the path of a biochemical clock. We also see how protein synthesis in another unicellular organism follows an all-or-none path characteristic of a dissipative structure.

In Chapter 12 the communication processes among the individuals in a colony of slime molds are shown to be another example of the self-organization of a biochemical medium. Messenger molecules in pacemaker cells spontaneously organize themselves into a collective mode of behavior and are delivered to the other members of the colony in a pulsating fashion.

The example in Chapter 13 deals with the physico-chemical laws which make it possible to generate complete and complex organisms, with many different cellular types, from a single fertilized egg. It is shown that the information for cell differentiation, in the form of a molecular gradient, is generated as a result of self-organizational processes inside the embryo. As the embryo grows, successive self-organized patterns arise and determine the shape and function of the organism.

Adult organisms themselves are the loci of many self-organizational processes. Circadian (diurnal) rhythms and regular heartbeats are two such processes. The brains of higher animals, particularly man, are certainly the ultimate in the self-organization of matter. Our thoughts, feelings, and actions result from the cooperative behavior of the ensemble of a few billion nerve cells. The modeling of an ensemble of intricately interconnected nerve cells raises new technical problems in the mathematical description of dissipative structures. In Chapter 14 these problems are illustrated by the example of the epileptic seizure, which is a pathological state of brain activity.

The general subject of the self-organization of matter has been treated in several books at various levels of complexity. *Thermodynamics of Structure, Stability, and Fluctuations* by P. Glansdorff and I. Prigogine, and *Self-organization in Nonequilibrium Systems* by G. Nicolis and I. Prigogine are advanced works. Both address themselves to researchers in the field. *From Being to Becoming* by I. Prigogine is at an intermediary level, whereas *Order Out of Chaos*, by the same author and I. Stengers, addresses the philosophical implications of self-organizing systems.

The Hypercycle by M. Eigen and P. Schuster and *Laws of the Game* by M. Eigen and R. Winkler are more oriented toward self-organizational problems in relation to the emergence of the genetic code. In *Advanced Synergetics*, H. Haken is more concerned with physical systems. He is especially interested in the origin of laser beams. Interested readers may find more specialized books referred to in the various chapters of the present work.

This book attempts to expound scientific processes which took innumerable research workers several centuries to discover. It is impossible to refer to all of those workers. I have chosen only to occasionally cite some famous names which probably are familiar to some readers. In the description of contemporary material I mention only researchers whose work has been used in the book.

Brussels, Belgium   A. BABLOYANTZ
January 1986

ACKNOWLEDGMENTS

My encounter with Professor I. Prigogine has been the turning point in my life. During more than two decades he has advised and guided me through the pathways of the theoretical study of nature. Starting from the study of the relatively simple hydrogen molecule, he has made me bifurcate by successive transitions into more and more complex areas of research. With his broad view of science and the humanities he has been a constant source of innovation and enthusiasm. I wish to thank him for his teaching, for making this book possible, and for his unwavering support on bad as well as good days.

My very special gratitude goes to Professor P. Glansdorff for spending so much of his time in reading and criticizing the manuscript. I have benefited greatly from his vast knowledge of thermodynamics and his precise mind in the preparation of the first part of this book.

I am also grateful to my friend and colleague Professor G. Nicolis for his constant support, helpful discussions, and his reading of the manuscript. His advice and encouragement have been invaluable not only for the preparation of this book but throughout my entire career.

The multidisciplinary nature of this book reflects the diversity of the subjects studied by friends and colleagues at the University of Brussels and the University of Texas in Austin. I wish to thank particularly P. Allen, P. Borckmans, K. Chemla, J. L. Deneubourg, G. Dewel, T. Erneux, C. George, A. Goldbeter, M. Herschkowitz–Kaufman, J. Hiernaux, L. Kaczmarek, D. Kondepudi, R. Lefever, M. Malek–Mansour, F. Mayne, A. Nazarea, C. Nicolis, S. Pahaut, M. Sanglier, A. Sanfeld–Steinchen, J. W. Turner, and D. Walgraef.

I am grateful to Professors B. Baranowski of the University of Warsaw, R. Mazo of the University of Oregon, R. Hansell of the University of Toronto, and V. Mathot, E. Vander Donckt and J. C. Legros of the University of Brussels for their criticisms of the manuscript.

It is a pleasure to thank C. Brian, J. McGill, and D. Newson for their help with English style.

The preparation of the manuscript was made possible with the efficient assistance of S. Dereumaux–Wellens, D. Hanquet, N. Sardo, and M. Adam. I thank them for their help.

I wish to express my especial gratitude to my friend P. Kinet, whose technical skill in preparing the illustrations for this book was invaluable.

A. B.

CONTENTS

INTRODUCTION

Man has always inquired about his roots. Usually he attributed his origins to supernatural forces or to a Divine Creator. In the nineteenth century Charles Darwin scandalized western civilization by stating that humans developed from inferior species via a long evolutionary process. In the mid-twentieth century Darwin's ideas were extended still further; humans, and indeed all living organisms, come from inanimate matter, and consist mainly of such ordinary substances as charcoal, oxygen, and hydrogen.

Although the molecular nature of living organisms has been indisputedly established for several decades, the physico-chemical laws that could "animate" molecules of inert matter into thriving life were not available until recently. With the advent of the theory of self-organizing systems, we now have a satisfactory understanding of how a purely physico-chemical system may complexify spontaneously into spatial and functional order. Today it is widely recognized that self-organizing properties play an important role in the unfolding of the processes of life. Such properties also give us some clues to the fundamental question of how life started. We are now in a position to construct plausible scenarios describing how, over billions of years, simple molecules evolved into complex organisms, following only the normal destiny imposed on them by the laws of nature. Our ideas about the origin of life are still in their infancy, but we believe that, whatever their future form, they will grow from the notion of self-organizing systems.

The three main parts of this book, which review such distinct subjects as chemical kinetics, thermodynamics, and molecular biology, may give this work the appearance of an encyclopedia. This reflects the fact that the subject of self-organization is the product of multidisciplinary research, which uses the language of mathematics, the evolutionary laws of thermodynamics, and new findings in the field of chemistry in order to explain the emergence of order in biological systems.

But before we start our multidisciplinary excursion into the study of the laws which govern the spontaneous organization of matter, let us look in more detail at the general direction of our journey.

*   *

*

The dawn of science goes back more than ten thousand years. In all early civilizations, humans performed chemical experiments, using fire to cook, melt metal, and make new substances not found in nature. They invented the first "machines" to lift and transport heavy loads, thus initiating a branch of science that today sends rockets into interplanetary space. Men of antiquity started the investigation and exploitation of nature by selecting plants as food or medicine, and by classifying birds, animals, and all other creatures into friend or foe.

In Western civilization it is customary to attribute the birth of science to the ancient Greeks. However, fragments of practical knowledge were known in Egypt, Chaldea, Phoenicia, China, and India. For example, the Egyptians used geometry for survey and construction purposes, and the beginnings of arithmetic may be found in Chaldea. However, the Greeks were the first people to embark upon the abstract study of science, both as a necessity as well as for the joy of the spirit.

Aristotle (384–322 B.C.) classified the sciences into two broad categories: theoretical (knowledge and speculation), and practical (action) sciences. The first group was in turn divided into three subgroups, physics, mathematics, and first philosophy.

Today, students of natural sciences may also be classified as experimentalists or theoreticians. The former probe nature with the help of whatever appropriate instruments are available. Experimentalists need a conceptual framework of theoretical considerations for the interpretation of their results. Theoreticians, on the other hand, incorporate experimental facts into the logic of mathematical methods. With such tools they can extend, by the so-called heuristic methods, the investigation of nature beyond what is visible and measurable. From a set of facts a theoretician constructs laws of increasing generality, but which ultimately must be confirmed by the experimental approach. Although frequent reference will be made to experimental data, our approach in this book is mainly theoretical.

*   *

*

At first, matter was investigated for its bulk properties, such as weight, consistency, or its affinity to combination with other substances. Gradually, regularities in the behavior of matter were expressed as laws. Soon it became evident that the same laws could be applied to small rocks as well as to heavenly bodies. Later the atomic nature of all matter was discovered. It was the nature of the interaction of these atoms that either gave cohesion to a piece of hard iron or transformed it into rust.

The twentieth century witnessed a drastic change in our views about the smallest part of matter. It turned out that atoms were made of a conglomerate of elementary particles and contained staggering forces in their nuclei. It was found that atoms contained particles with size of the order of 10^{-13} cm and objects that had a lifetime of 10^{-22} second.

In the beginning the investigation of living organisms was only descriptive. Later, the study of various functions of organisms revealed the physico-chemical nature of vital processes. For example, it was found that respiration was a form of combustion, digestion involved chemical reactions, and blood circulation could be related to ordinary fluid flow. Thus a first link was established between physics, chemistry, and living organisms.

In the nineteenth century it became obvious that all living matter is made of collections of units called cells. Later, the living cell itself turned out to be a highly complex organized structure. For example, a cell contains assembly lines for the synthesis of proteins, information storage units that ensure reproduction, and little organelles that support respiration. The great advances in biochemistry during the mid-twentieth century showed that the cell is a chemical "factory" and its "vital forces" merely a set of ordinary physico-chemical transformations. The information blueprint for these biochemical processes, and for the reproduction of cells, is coded with the help of a four-letter molecular alphabet. In the final analysis, it seems that living organisms may be considered and analyzed as assemblies of molecules and atoms in dynamic interaction.

In the physical sciences, matter is studied at several levels of organization. A great deal of information is now available about the internal structure of atoms, their motion, and their mutual interaction to produce more complex entities. All this knowledge constitutes the *microscopic* view of nature. On the other hand, in a *macroscopic* view of matter, theories quantify the properties of bulk matter—predict, for example, its electrical or magnetic properties, or the affinity of substances toward each other. All these theories have been extremely successful in taming nature, making possible today's advanced technological society.

One of the macroscopic laws of nature indicates the direction of change of properties of bulk matter. It tells us, for example, why a drop of ink diffuses immediately in a glass of water instead of remaining as a localized blue spot. Or it predicts that heat always flows from the hot regions into the cold areas. In such cases we speak of *irreversible processes*: once they have happened, the opposite process becomes impossible. The dissolved ink will never spontaneously form a small drop again.

We predict the direction of evolution of an irreversible process with the help of a law which introduces a quantity called *entropy*. The law states that, in the presence of irreversible processes, the entropy of an isolated ensemble of matter, called a *system*, increases and reaches a final maximum value. At this value of entropy, all possibility of change disappears, and one speaks of matter in an *equilibrium* state.

The increase in entropy may be associated with an increase in disorder in the relative arrangement of molecules. For example, ice cubes in a glass of whiskey show an orderly segregation of cold water molecules from the warm alcohol of the whiskey. As the ice melts, this order disappears and a homogeneous mixture of uniform temperature is obtained. This process cannot be reversed as long as the glass remains isolated from the surroundings.

In the middle of the nineteenth century, considering the ensemble of irreversible processes in isolated systems, the German physicist Rudolf Clausius wrote: "The entropy of the Universe tends to a maximum value." This implied that the tendency of the universe is toward increasing disorder, to a state of equilibrium, in which all future possibilities of change are eliminated.

Fortunately, experience shows that this is not the case in the world of living organisms. On the contrary, during the billion years of evolution, in the course of a long journey toward complexity the human species was gradually produced from a single cell. Within a few years a tiny almond seed apparently made of a homogeneous white substance grows into a tree. In living organisms spatial and functional order is the rule. For example, the regular beat of the heart, a functional order, defies the tendency toward maximum entropy.

How this can be? Why do inert matter and living species evolve in opposite directions if both are descendants of the same atom? The answer to these questions comes from the realization that if it is true that the entropy of a large quantity of matter taken as an isolated entity must increase, nevertheless the local entropy of a fraction of that matter may in some cases decrease, thus generating local order at the expense of the neighboring parts.

In order to reconcile matter and life, the macroscopic laws of the nineteenth century had to be reconsidered and extended to such open systems. It was found that, under well-defined conditions, matter may self-organize autonomously into a lower entropy and more ordered state. Moreover, it turned out that a good part of the biochemical reactions inside organisms are good candidates for self-organization.

The importance of the macroscopic theories of self-organization of matter was underlined by the fact that, at a time when scientific research had gone in the direction of the microscopic view of matter and into the depth of the atom, the Nobel Prize for chemistry was awarded to L. Onsager (1965) for building a bridge between the microscopic and macroscopic view of irreversible processes. The same prize was given to I. Prigogine (1977) for his work in the field of self-organizing irreversible processes.

The spontaneous emergence of temporal and spatial organization, collectively termed "structures," can only arise and be maintained by a constant exchange of energy and matter between the self-organized ensemble and the external medium. For this reason they were named by Prigogine *dissipative structures*.

These extremely important advances in a theory which predicts the self-organization of matter into temporal and spatial order still had to be confirmed by direct experiments. Scattered results from various fields were available, but they went unnoticed or were considered as artifacts and were ignored by the great majority of researchers, for lack of theoretical support. As soon as there was a conceptual framework for the understanding of the self-organization of matter, there was renewed interest in the study of this topic. Soon the various organized states of matter could be traced to a common origin.

One of the most spectacular examples came from the field of chemistry and revolutionized our view of reacting mixtures. For centuries, chemists had been accustomed to mixing various ingredients which reacted together more or less rapidly and produced new substances. For example, if all reacting chemicals were soluble in water and the product of reaction was red, all one could see would be a gradual coloring of the medium into a final and permanent red solution, motionless, uniform, dull and with no memory of the past. In other words, the system reached an equilibrium state, and we were in the presence of a *static* chemistry. This behavior was predictable and compatible with the law of evolution toward the maximum entropy, as stated above.

Today, with the discovery of oscillating reactions we can speak of a new chemistry, one of motion, and of change, a *dynamic* chemistry. For example, in the well-known Belousov-Zhabotinski reaction one mixes into a flask four quite ordinary, water-soluble chemicals, some of them known for centuries. According to the relative amounts of the ingredients, the most spectacular results can be seen. In some instances the reaction medium behaves as a chemical clock. The fluid changes color at regular intervals, measuring the inexorable unfolding of time. The machinery of this clock is formed solely from molecular interactions.

Under slightly different conditions this same reaction may give rise to horizontal layers of red and blue stripes, thus organizing the system into a spatio-temporal structure in contradiction to the tendency of systems toward maximum disorder. Under other reaction conditions, a thin layer of the reacting mixture gives rise to dynamic works of art. Red and blue concentric curves gradually unfold in the medium around several centers and generate abstract dynamic patterns. The Belousov-Zhabotinski reaction also displays memory-like properties. It can discriminate between past and present, or exhibit an erratic behavior.

In classical chemistry, the fate of bulk matter is determined by almost instantaneous interactions of few molecules. In the new chemistry, it is the large-scale cooperation of all molecules in the flask that generates spatial and temporal order. Moreover, the reaction product feeds back on the chemical process and produces a snowball or catalytic effect. Thus, one might say that dissipative structures are characterized by "openness," "feedback," and "cooperativity." A proper understanding of the cooperative self-organization of matter requires a dynamic approach which can be achieved only by a theoretical treatment that generally requires the use of mathematical formalism.

The reason for such a late discovery of systems that self-organize spontaneously is that only reactions with well-defined properties can exhibit such a seemingly extravagant behavior. Several laboratories are now actively engaged in the discovery of such reactions, the number of which is steadily increasing.

We have seen that biochemists study living organisms by reducing them to purely physico-chemical systems. On the other hand, experience shows that the conditions for the onset of self-organized chemical reactions are satisfied by most biochemical processes inside living organisms. In these cases the reaction media are open to inflow and outflow of matter; reaction products feedback and enhance their own production, thus generating a snowball effect. Therefore, it is natural to think that dissipative structures are the basic processes that underlie various important structural and functional aspects of living organisms. This fact at first sight seems to contradict the maximum entropy law.

Another crucial aspect of the theory of self-organizing systems is that it is currently the only framework in which one may hope to explain, with the help of physico-chemical laws, how life emerged from inert matter. Granted that the onset of self-organization is imposed by physico-chemical laws, there is an infinite variety of dissipative structures that could have emerged. The specific choice of "our" life may be seen as the result of a long evolutionary process through which self-organizing systems passed.

On our own planet, theories of dissipative structures root man securely in nature. Man becomes an integral part of his environment, and is no longer unique in the immensity of the universe. Even his formidable brain stems from and functions according to the same ordinary laws which govern inert matter. We may conclude by stating that dissipative structures, which arise from the cooperation of physical laws and random events, appear to be the missing link between inert matter and life.

There are other instances in which the properties of an ensemble cannot be guessed from the behavior of its components. If feedback mechanisms and a permanent interaction with the environment exist for such ensembles, then these systems obey the same laws and logic as the molecules of the new chemistry. Therefore it is not surprising to see that today the concepts and methods of dissipative structures have entered not only other fields of natural sciences but are used in the social sciences as well. Who can guess the behavior of a society and its economic state from the observation of one individual and his checking account?

Part I

MATTER AND CHEMISTRY

How and of what is the world made?

Thales of Miletus Sixth Century B.C.

Chapter 1

THE STRUCTURE OF MATTER

1.1. MAN AND MATTER: A RELATIONSHIP OF SEVERAL MILLENNIA

All ancient civilizations used practical or artisanal chemistry; their craftsmen extracted and blended metals and made glass. Metallurgy on an industrial scale was developed in the mountains of Armenia some time between the twelfth and tenth century B.C. Chemical industries flourished in Mesopotamia, where craftspersons produced pottery, glass, glaze, paints, dyes, cosmetics, perfumes, and beer. The Mesopotamians recorded one of the oldest chemical texts (seventeenth century B.C.) in the form of a small cuneiform tablet conserved in the British Museum. It describes the making of glaze from copper, lead, clay, and verdigris. Witchcraft and medicine based on the chemical properties of herbs were also practiced, even in very early and primitive societies.

With the prosperity and progress of societies came the time of inquiry. What is matter? From where did its endless diversity come? Every civilization had its own answer to these questions. According to one line of thought, the gods put order into chaos and created matter; in another version, the gods created matter out of nothingness. The early Greeks thought that matter was made from a unique "primal" constituent that could take on many different aspects. Later on, Leucippus invented the atomic theory, which was developed further by Democritus. This theory states that the world is made of the empty and the full. The fullness is divided into an infinite number of small atoms that are simple, indivisible, and eternal. Atoms differ only in shape, order, and position. They may combine in an infinity of ways, thus generating an endless variety of substances and forms.

In the time of Aristotle it was thought that all matter was formed by a combination of "elements." Aristotle distinguished one celestial and four terrestrial elements. The terrestrial elements were earth, water, air, and fire, as seen in Fig. 1.1, and they conveyed the sensations of coldness, heat, wetness, and dryness. Terrestrial elements moved in rectilinear motion, whereas the celestial ones had a circular motion.

In the Middle Ages this notion was extended considerably (as seen in Fig. 1.2) and formed the philosophical cornerstone of that period of civiliz-

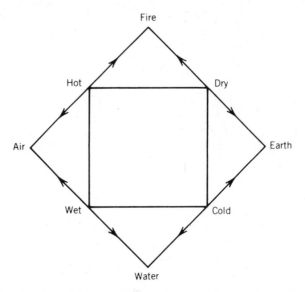

Fig. 1.1. Aristotelian terrestrial elements. (From I. B. Hart, *Makers of Science*, 1923, reprinted by permission, Oxford University Press.)

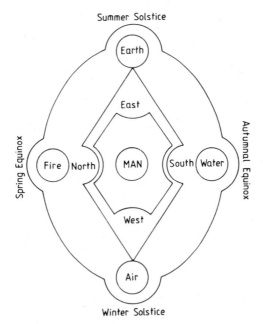

Fig. 1.2. Microcosm and Macrocosm. The tenth century view of the universe. (From I. B. Hart, *Makers of Science*, 1923, reprinted by permission, Oxford University Press.)

10

ation: man (microcosm) and the world around him (macrocosm) were seen as interrelated; the microcosm reflects the macrocosm, thus determining the fate of the individual. Astrology was born from such considerations, and continues to thrive in our super-scientific, atomic age.

The Middle Ages were also the era of the western alchemists. (A comparable mystical approach to matter had also flourished in China and India in much earlier times.) The shining gold particles in river beds have attracted humans since early times. This metal had mostly symbolic and ornamental value, but even in this century we used the gold standard in our economic exchanges. For alchemists, gold was unalterable, perfect, and eternal, virtues that all men must pursue. Therefore, it is not surprising that this metal became an important element in their metaphysical concepts. Given enough time, the final destiny of all imperfect metals was the eternal gold, and that of man, immortality. But alchemists thought they could accelerate the temporal rhythm of the cosmos and bring about the transmutation of man and metals into their perfect states, if they could find the philosopher's stone. This imaginary entity was sought intensively in the "laboratories' of the Middle Ages. Alchemists thus created a chemical technology, discovered new products, and synthesized the "elixir" of longevity.

Unfortunately, they kept their findings secret; chemistry had to wait until the seventeenth century to be recognized as an autonomous discipline. Then chemical knowledge became public and was recorded in common and clear language. Paracelsus introduced the idea of five "principles" responsible for the progress of a chemical process. To the Aristotelian elements of water and earth he added the principles of mercury, salt, and sulphur, present in different quantities in all matter, and determining their reactivity. Boyle (1627–1691) introduced a more modern idea of the element, which he defined as the ultimate and nondecomposable subdivision of all matter.

The eighteenth century was a turning point for chemical knowledge. The existence of the gaseous state of matter was discovered, thus providing an explanation to many of the unsolved problems that faced the chemists of that time. Chemistry emerged as an autonomous science, the youngest of all sciences, and following the older branches of knowledge adopted Cartesianism. The notions of "caloric" and "phlogiston" are also products of eighteenth century thinking. Every combustible material had a principle, called *phlogiston*, which was liberated by burning the material. Heat was formed from indestructible material particles, called *calorics*. The French scientist Lavoisier (1743–1794) is considered the father of modern chemistry. By careful and rigorous experimentation, Lavoisier refuted the existence of phlogiston. He showed that all combustion processes were just chemical reactions. For example, the respiration of living beings was combustion without flames.

At the beginning of the nineteenth century, Dalton revived and modernized the Greek atomic theory. In order to explain the laws of the combination of substances, he postulated that simple matter is formed from indestructible atoms, like billiard balls; all existing materials are combinations of atoms of various kinds. Throughout the nineteenth century, following the other sciences, chemistry adopted and used mathematics extensively in establishing laws and a rational framework for the study of chemical processes. By applying thermodynamic laws (see Chapter 3) to chemistry, a link was made between physics and chemistry. This interpenetration was to continue with the kinetic theory of gases. The nineteenth century also saw the birth of the chemical industry. Thousands of new products were synthesized, the kitchen-like alchemists' laboratories gave way to gigantic industrial complexes, and chemistry became an all-important branch of activity among industrialized nations.

The first decades of the twentieth century witnessed a great leap forward in our understanding of the physical world. The quantum theory of atoms revolutionized the field of physics, and chemistry had to follow. At last chemists could understand why substances react with one another. Since then, chemistry has become more and more dependent on physical concepts and complex mathematical notions. In twenty-five centuries, the notion of matter had evolved from that of tangible objects to that of elementary particles buried in the depths of an atom. However, a purely reductionist view of chemistry cannot explain the unexpected phenomena that were encountered in some chemical reactions. In the mid-fifties a new chemistry had to be elaborated. It developed from the macroscopic concepts of the last century. This new chemistry will be the subject of the rest of the chapters of this book. But first we shall summarize a few concepts of classical chemistry.

1.2. ATOMS

By the end of the nineteenth century, the idea of the indestructible atom was on its way out. Mendeleev constructed his periodic table of the elements, which presently has been extended to more than 108 atoms. J. J. Thomson had discovered small particles called *electrons* inside the atom considered as an indestructible billiard ball. At the same time, the discovery of natural radioactivity by Becquerel and the Curies, and the existence of spectral lines, pointed to the fact that atoms are very complex structures. In 1911 Rutherford discovered the existence of the atomic nucleus; immediately afterward, Rutherford and Bohr proposed their planetary model of the atom, formed of a central nucleus with electrons gravitating around it in well-defined orbits (see Fig. 1.3). In 1919, the first artificial transmutation of atoms was performed, and modern physicists made the ancient alchemists' dream come true.

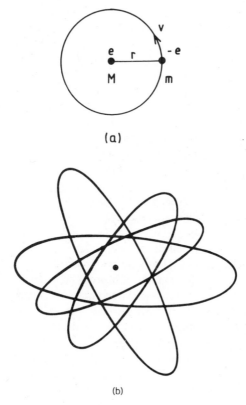

(a)

(b)

Fig. 1.3. (a) Bohr's model of the hydrogen atom. A single electron with charge − e revolves in a circular orbit around the heavy nucleus, with charge + e and assumed to be at rest. (b) A multi-electron atom.

However, the spirit was radically different. Since Descartes, a scientist's task was not only to help the processes of nature but to dominate and manipulate the natural elements for his own benefit.

It was soon realized that nuclei themselves are not simple but are made of a great number of small particles, such as protons and neutrons. Between 1940 and 1970, many *elementary particles* in addition to protons and neutrons were discovered. Today, we believe that all matter is formed from the same *fundamental particles*, which are divided into two families of six *leptons* and six *quarks*. The electron is a lepton, whereas protons and neutrons are made of assemblages of quarks.

Four different types of forces operate in nature. The *gravitational* force, discovered by Newton, has been recognized for the longest time and in the most intuitive fashion. *Electromagnetic forces* operate between entities endowed with *positive* and *negative* electric charges, and act through the intermediary of particles of light called *photons*. To these forces we owe the development of all our electrical gadgets. The two other forces in nature act between elementary particles: *weak forces* act both on leptons and quarks, while *strong forces* operate only between quarks. These forces keep the elementary particles in a small tightly assembled conglomerate which forms the *nucleus* of the atom. The unlashing of these forces is the source of the destructive power of atomic bombs; when harnessed, these forces provide the energy to run nuclear power plants.

A particular atom is characterized by a well-defined number of elementary particles (see Fig. 1.4) and by the number of electrons that surround the nucleus. However, as our aim in the present chapter is only to understand the emergence of chemical species and their mutual interactions, a few fundamental particles will suffice. These are neutrons, positively charged protons, and negatively charged electrons (which have a mass of 1/1836th that of the neutron). Because of the positive charge of protons, the nuclei are positively charged. The quantity of positive electricity in a nucleus is directly proportional to the number of protons of which it is composed. The light, negatively charged electrons move around the nucleus according to well-defined rules, and are attracted by it because of the action of electromagnetic forces. Their number per atom is determined by the number of protons, in such a way as to balance the positive charge of the nucleus.

The size of an atom is determined by the distance between electrons and the

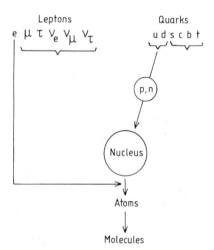

Fig. 1.4. All matter is formed from the same elementary particles: u = up, d = down, c = charm, s = strange, b = bottom, t = top, μ = muon, τ = tau, v = neutrino, p = proton, n = neutron, and e = electron. Ordinary matter contains only p, n, and e.

nucleus. The nucleus accounts for most of the mass of the atom. The lightest known atom, hydrogen, weights 1.67×10^{-24} grams and has a diameter of 0.5×10^{-8} centimeters. The concept of the atom is not just a convenient theoretical hypothesis; large molecules and their constituent atoms can be detected by the electron microscope (see Fig. 1.5), while the elementary particles may be detected by the tracks they make in appropriate experimental chambers, as seen in Fig. 1.6.

Today there are 108 different known atoms. Arranged according to their increasing mass, as seen in Fig. 1.7. A cohesive set of atoms of the same kind form *pure substances* which have distinct characteristic physical and chemical properties. The well-known substances, such as iron, gold, silver, and copper, are, of course, part of this table. However, chemists customarily designate substances by their Latin names. Moreover, each atom has a symbol, constructed from the first or the first two letters of its Latin name; for example, Fe stands for Ferrum (iron) and Au for Aurum (gold). If atoms of a given element differ only by the mass of their nuclei, they are called *isotopes* (same

Fig. 1.5. This photograph taken by electron microscope shows the regular arrangement of molecules (hexadecachlorophthalocyanate of copper) and their constituent atoms. (Courtesy of N. Uyeda.)

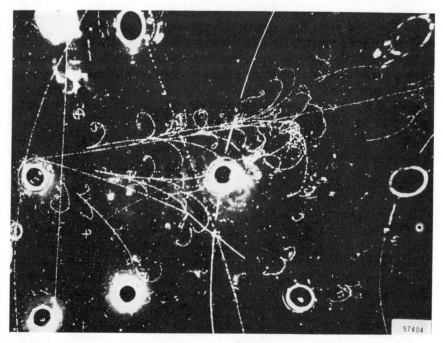

Fig. 1.6. Elementary particles are detected by tracks they make in appropriate experimental chambers. (Photo CERN.)

number of protons, different number of neutrons). In the periodic table the upper figure gives the number of electrons. The mass of one carbon isotope is set arbitrarily at 12,00000, and the atomic masses of other elements are determined relative to this value. Thus the hydrogen atom appears to have a mass approximately 1/12 that of the carbon atom.

1.3. MOLECULES

Atoms combine and form molecules, which are more stable structures than the original atoms. The degree of eagerness of an atom to form molecules (its reactivity) is a function of its electronic configuration. Some atoms such as oxygen, O, and hydrogen, H, are so reactive that, even in the absence of other substances, they combine with an atom of their own species and form hydrogen or oxygen molecules, denoted respectively by H_2 and O_2. (When a subscript, always an integer, follows the symbol of a given atom in a complex molecule, it indicates the number of atoms involved in that molecule.) In

1 H																	2 He
3 Li	4 Be											5 B	6 C	7 N	8 O	9 F	10 Ne
11 Na	12 Mg											13 Al	14 Si	15 P	16 S	17 Cl	18 Ar
19 K	20 Ca	21 Sc	22 Ti	23 V	24 Cr	25 Mn	26 Fe	27 Co	28 Ni	29 Cu	30 Zn	31 Ga	32 Ge	33 As	34 Se	35 Br	36 Kr
37 Rb	38 Sr	39 Y	40 Zr	41 Nb	42 Mo	43 Tc	44 Ru	45 Rh	46 Pd	47 Ag	48 Cd	49 In	50 Sn	51 Sb	52 Te	53 I	54 Xe
55 Cs	56 Ba	57 La	72 Hf	73 Ta	74 W	75 Re	76 Os	77 Ir	78 Pt	79 Au	80 Hg	81 Ti	82 Pb	83 Bi	84 Po	85 At	86 Rn
87 Fr	88 Ra	89 Ac	104 Unq	105 Unp	106 Unh												

58 Ce	59 Pr	60 Nd	61 Pm	62 Sm	63 Eu	64 Gd	65 Tb	66 Dy	67 Ho	68 Er	69 Tm	70 Yb	71 Lu
90 Th	91 Pa	92 U	93 Np	94 Pu	95 Am	96 Cm	97 Bk	98 Cf	99 Es	100 Fm	101 Md	102 No	103 Lr

Fig. 1.7. Periodic table of elements. Only 103 elements are shown.

ordinary circumstances atoms such as gold, Au, have a very low reactivity; a fact that explains the inalterability of gold.

Oxygen and hydrogen combine to produce the all-abundant and vital water molecule, H_2O. In this case, for each atom of oxygen, two atoms of hydrogen are required. More generally, atoms enter a molecular structure in varying but well defined numbers. This fact may be seen at the macroscopic level by the observation that different elements always combine together in a chemical reaction in amounts that are the ratios of small integers. Precisely these observations led Dalton in the nineteenth century to revive the old atomic theory. Dalton's original view of molecules was different from the Greek atom in the sense that every element was associated with its constituent atoms.

Molecules themselves may combine to form still more complex molecular structures. Sometimes they may reach gigantic dimensions in biological organisms, as we shall see in Chapter 10. The notation used in chemistry for complex molecules is as follows. One writes all the atomic symbols of the constituents of the molecule in succession, each atom bearing a subscript showing the number of times it is present in the molecule. For exemple, ethanol, the alcohol in whiskey, is designated by C_2H_5OH. Sometimes as in the case of ethanol, some atoms, such as the H in OH, are distinguished from other atoms of the same kind, such as the H of H_5. This is a convenient way to indicate that these atoms are in a different environment in the molecule.

1.4.   QUANTUM MECHANICS

The question of how and why neutrons, protons, and electrons form atoms and molecules is studied today in a branch of physics called *quantum mechanics*. Quantum mechanics is one of the major breakthroughs of twentieth century scientific thinking. It was elaborated by such well-known scientists as Planck, Einstein, de Broglie, Schrödinger, Heisenberg, and many others.

Until the beginning of this century, the study of nature was approached by two essentially different concepts. The idea of continuously propagating waves was used to explain electromagnetic phenomena, such as the propagation of light rays. On the other hand, the particle nature of atoms and molecules made them suitable for study in the framework of the laws derived from classical Newtonian mechanics. However, in the late nineteenth century, with the refinements of experimental techniques, many unexpected and unexplained phenomena were revealed which forced the scientific community into a revolutionary way of thinking. De Broglie made the dazzling hypothesis that particles and waves are two ways by which matter can manifest itself. In other words, to each material particle in motion is associated a wave having a wavelength inversely proportional to its mass and velocity. We are heavy and are not aware of the wave that is trailing us, but it gives extraordinary properties to the lightweight electrons. A beam of electrons under the proper experimental conditions acts like a train of light waves.

One of the major features of quantum theory is Heisenberg's well-known *principle of uncertainty*. According to this principle, it is impossible to know simultaneously the exact position and momentum of a particle. The product of estimated error on both quantities must always remain larger than a well-defined constant value. This is contrary to classical mechanics, where, in principle, both quantities may be determined with any desired accuracy. In fact, as electrons revolve around the nucleus, we can only evaluate what is the chance (probability of presence) of finding one at a given point and at a given time. Again, had they followed trajectories (orbits), as was assumed in the early, planetary models of the atom, their position and momentum could be determined exactly, in violation of the uncertainty principle.

This modern view of the atom, illustrated for the simplest of all atoms, the hydrogen atom, is shown in Fig. 1.8. The number of small dots at a given point represents the *probability density* of finding the unique electron of the atom at that point. More dots in a given region are an indication of a better chance of finding the electron in that region. At some distance from the nucleus, the probability of presence of electrons drops sharply, thus determining the volume of the atom. The probability density for hydrogen and other more complex atoms is computed by solving a partial differential equation called *Schrödinger's* wave equation, which incorporates ideas and mathematical formalism beyond the scope of this book.

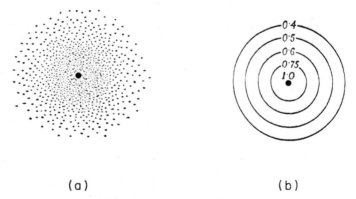

(a)   (b)

Fig. 1.8. A modern view of the hydrogen atom. (a) The density of small dots shows the probability of finding the unique electron around the positively charged nucleus. (b) Another visualization: a contour map of electron probability density for the hydrogen atom. (From C. A. Coulson, *Valence*, 1952, reprinted by permission, Oxford University Press.)

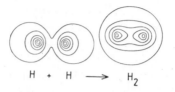

Fig. 1.9.   Formation of hydrogen molecule. (From C. A. Coulson, *Valence*, 1952, reprinted by permission, Oxford University Press.)

In a liquid or gaseous phase, atoms are endowed with kinetic energy, they are constantly in motion, and may reach the immediate vicinity of other atoms. For example, when two hydrogen atoms approach each other (see Fig. 1.9), the electron of atom one is attracted by the proton in the nucleus of atom two, while the proton in the nucleus of atom one attracts the electron of atom number two. At the same time, repulsive forces act between two negatively-charged electrons and two positively-charged nuclei. Finally, all the attractive and repulsive forces balance each other and due to quantum effects a stable hydrogen molecule, H_2, is formed. The latter is much more stable than the individual atoms forming it. Now, if a third atom H approaches the molecule H_2, it cannot form a three-atomic hydrogen molecule H_3. The same type of reasoning holds for any complex molecule.

The quantum effects result from the fact that electrons possess an intrinsic property called *spin*. If a pair of electrons are considered, it is customary to designate them as *parallel spins* ($\uparrow \uparrow$) if they have the same value and to call them *antiparallel* ($\uparrow \downarrow$) if they are of opposite values. Two electrons of parallel spin avoid each other, whereas two electrons of antiparallel spin have a tendency to pair with each other and to discriminate against any additional electron. The possibility of atoms assembling into molecules depends essentially on the spin and the configuration of electrons. A hydrogen molecule can only be formed if two atoms come together with antiparallel spins. H_3 is impossible because the single atomic electrons are already paired in H_2.

1.5.   CHEMICAL BONDS

Following the same line of reasoning, one can also explain the formation of complex molecules. For example, the nucleus of a nitrogen atom, one of the components of the ambient air, is formed from seven neutrons, seven positively charged protons. Seven electrons move around the nucleus. The probability density of the electrons around the nucleus is not uniform, but is distributed in various layers and sublayers. According to the laws of quantum mechanics, each layer may only contain a well-defined number of electrons. Atoms such as argon, with full layers, are ordinarily nonreactive. Only those atoms given in the last column of Fig. 1.7 have such an ideal electronic distribution. The remaining atoms must tend toward this ideal by combining and sharing electrons with other incompletely filled atoms. The number of electrons a given atom may share with other atoms determines the maximum number of bonds that the atom is able to form. This number is called the *valence* of the atom.

For example, the carbon atom, C, has a maximum of four valence electrons. In order to complete a layer and form an octet (this number is determined by the Schrödinger wave equation), it needs four extra electrons. Thus, C can react with four monovalent atoms in the combustion gas methane, denoted by chemists as CH_4, or

$$\begin{array}{c} H \\ | \\ H-C-H. \\ | \\ H \end{array}$$

(In the second notation each bar designates one chemical bond.)

With every exhalation, we eject carbon dioxide, CO_2, into the atmosphere. This is the waste product of combustion processes that take place in all living organisms. In this molecule, each oxygen atom needs two electrons for the formation of an octet. We say a double bond is formed between the carbon and oxygen atoms and we note $O=C=0$. In the hydrogen molecule of Fig. 1.9, the probability of finding the paired electrons around both nuclei is the same everywhere; this type of association forms *covalent bonds*.

A sodium atom Na and a chlorine atom Cl, may form a sodium chloride molecule, NaCl, which is ordinary table salt. In this molecule, the paired electrons responsible for bond formation have a greater probability of being found around the Cl nucleus, which accumulates a small negative charge. Consequently, the sodium atom gains an equivalent amount of positive charge. However, the molecule as a whole remains neutral. This type of bonding is called *ionic*.

Atoms or molecules can have one or several electrons stripped from their periphery, thus becoming positively charged; they are called *positive ions* (or cations). For example, a hydrogen molecule may still exist with only one electron circling both nuclei. The notation is the same as for atoms and molecules, but one or several + signs indicate the number of missing electrons. H_2^+ and Cu^{2+} are respectively a single-charged hydrogen molecule and a double-charged copper atom. If a hydrogen atom is stripped of its unique electron, the resulting H^+ ion is a naked nucleus, made of one proton.

In other instances, molecules and atoms may gain extra electrons and become negatively charged. We say they are *negative ions* (anions). When we dissolve table salt in water, we are left with only Na^+ and Cl^- ions surrounded by water molecules. Normal water itself contains a small quantity of OH^- and H^+ ions. When the concentration of protons in a solution is in excess of their value in pure water, the solution is an *acid*. Ordinary vinegar is an example of an acid solution. An excess of OH^- ions in a solution defines a *base*.

1.6. CHEMICAL REACTIONS

Chemical reactions that occur as part of the natural processes of life, and those performed in laboratories and factories to yield products unknown to nature, are usually extremely complex. Chemists need a "shorthand" language in order to represent these reactions, and at the same time to convey the maximum amount of information about them. We illustrate this point with the example of the reaction of chlorine with hydrogen molecules. At the molecular level, the succession of events is illustrated in Fig. 1.10 and the set of these transformations is called the *reaction mechanism*. In chemists' notation, this reaction reads simply

$$Cl + H_2 = ClH + H \tag{1.1}$$

and is known as a *stoichiometric equation*.

Unfortunately, all chemical processes cannot be represented in such a simple way. For example, we all know that hydrogen and oxygen molecules combine together to form water. If we write $H_2 + O_2 = H_2O$ and try to represent it with a process such as that of Fig. 1.10, we are left with an extra oxygen atom. No experimental evidence of such an isolated oxygen atom has been found in the reaction vessel. However, a formula such as $2H_2 + O_2 = 2H_2O$ represents adequately the experimental data.

In general, reactants A, B, C,... may combine in various well-defined proportions to give products G, H,... according to the stoichiometric

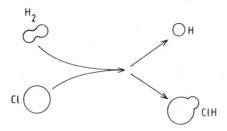

Fig. 1.10.   The one-step collision of Cl and H_2 produces H and ClH.

equation

$$vA + \mu B + \eta C + \cdots = \gamma G + \delta H + \cdots \qquad (1.2)$$

The coefficients v, μ, η, γ, and δ are called *stoichiometric coefficients*.
A simple stoichiometric equation such as

$$H_2 + Br_2 = 2HBr$$

which, at first sight, seems analogous to the reaction given in Eq. (1.1), does not necessarily guarantee a simple mechanism. As we shall see later, chemists have experimental techniques for finding out how chemical bonds are formed. In the case of this reaction, the formation of HBr requires three steps instead of the single reaction process shown in Fig. 1.10. In simplified notation, the mechanism and the *global reaction* of the formation of HBr is written as

$$Br_2 \rightleftharpoons 2Br$$
$$Br + H_2 \rightarrow HBr + H$$
$$H + Br \rightarrow HBr \qquad (1.3)$$

Thus

$$H_2 + Br_2 = 2HBr$$

In the last two steps, an arrow has been substituted for the equality sign in order to show the direction of progress of the chemical reactions (*direct reactions*). Since the newly formed bromine atoms may combine to form Br_2 again, two arrows are needed in the first step to indicate the possibility of *reverse reaction*. Substances such as H and Br that appear and disappear without appearing in the global reaction are called *intermediate components*.

This example shows that in general the stoichiometric equation indicates only the overall reaction process. The molecular details of the reaction are usually much more intricate than is suggested by the stoichiometric equation. A complex mechanism may involve ions, atoms, molecules, and *free radicals*. The latter are ordinary atoms or molecules, but are endowed with high reactivity.

1.7. STATES OF MATTER

Matter at the macroscopic level, as seen by the naked eye, may contain more than 10^{23} atoms or molecules. If all atoms or molecules are of the same kind, we have a *pure* substance (for example, carefully distilled water); otherwise we deal with a *mixture*. A mixture or a pure substance may exist under different disguises. For example, water freezes at 0° and turns into steam at 100°C, at one atm. illustrating the liquid, solid, and gaseous "phases" of matter. More broadly, a *phase* is defined as a uniform portion of the matter where the physical and chemical properties are everywhere alike. A glass of pure whiskey is a one-phase mixture and also a *homogeneous* system; a glass of whiskey on the rocks is a two-phase mixture and is *heterogeneous*.

*   *

*

Chemical reactions play a very important part in the processes of self-organization of matter, especially in relation to the origin and present-day organization of living entities. As we shall see later, the investigation of such phenomena necessitates a dynamic approach to chemical processes. For this reason, in the next chapter we introduce, with some details, the important elements of chemical kinetics.

Chapter 2

CHEMICAL KINETICS

2.1. MACROSCOPIC AND MICROSCOPIC VIEWS OF MATTER

A human body is an assembly of cells. The cells are formed from molecules, which in turn are combinations of atoms and, consequently, conglomerates of elementary particles. The odyssey from the elementary particles to man is a long one, and, granted the present state of knowledge, it cannot be traveled in a single, continuous stretch. However, here and there we have a satisfactory knowledge of some of the processes that lead from inanimate matter to animated nature.

As we go down the road jumping from one known area to the next, we assume the existence of entities which were precisely the subject matter of the study of a previous domain. For example, by assuming the existence of atoms, with their positively charged nuclei and negatively charged electrons, we understand the mechanism of formation of hydrogen molecules. However, the structure of atoms and the nature of the cohesive forces which act between particles inside nuclei must also be investigated.

Our knowledge of physics may be divided into *microscopic* and *macroscopic* sciences. Each branch in turn subdivides into several subgroups—the nature and the laws of interaction of elementary particles, the electronic structure of atoms and molecules, and the various modes of interaction of these particles in bulk matter form the essence of the microscopic sciences. Macroscopic descriptions of nature, on the other hand, account only for the bulk properties of matter and ignore completely its atomic nature. For example, one studies the laws governing the propagation of heat in a piece of metal, or the flow of electricity in a long wire. The technology of all photographic equipment and eyeglasses derives from the macroscopic science of optics.

The macroscopic approach to the study of nature preceded microscopic description. Its aim was to account for the measurable and tangible properties of matter accessible to experimentation with the technology of the eighteenth and nineteenth centuries. It was quickly subdivided into different branches: mechanics, thermodynamics, optics, electricity, magnetism, and chemistry.

At the macroscopic level of investigation a chemist is interested in the following problems. He must know if a given mixture of chemicals will react and produce new products. He must be able to compute the yield of such a reaction and determine all the factors that influence the formation of the new products. All these questions constitute the subject matter of *chemical thermodynamics*. The salient features of this macroscopic science will be sketched in Chapter 3.

All chemical reactions proceed at a definite rate. Therefore the chemist must determine all the factors which influence these rates, and be able to act upon them in order to accelerate or decelerate the formation of the products of the reaction. These problems constitute the subject matter of *chemical kinetics*, which will be reviewed in the present chapter. We have developed this chapter in some detail, since the processes of self-organization of matter are of a dynamic nature and rely on the science of chemical kinetics. Chemical kinetics was developed in several laboratories in the western world during the first half of the twentieth century. This relatively young branch of chemistry is the work of many researchers; its history cannot be told in a few sentences.

2.2. CHEMICAL KINETICS

The word *kinetic* means "pertaining to motion," thus indicating the idea of a dynamic view of reaction processes. A lit match provokes an immediate explosion in a room full of inflammable gas, whereas an iron nail needs several months to become rusted completely. In both cases we are dealing with chemical reactions, but with two different rates. Chemical kinetics is the study of these rates and their relevance to the elucidation of the reaction mechanisms.

Our first aim must be to define the *rate* of a chemical reaction. Intuition tells us that the rate must be related to the change in the number of new molecules that are formed or reactants that disappear per unit volume and unit time. We want this number to be independent of the size of the system. Thus chemical kinetics is a macroscopic approach, in contrast to a microscopic description of single-molecular events as described in Fig. 1.10.

One of the units for the evaluation of the number of molecules in a unit volume (liter), is the *mole*. An ensemble of $N = 6.02 \times 10^{23}$ molecules, called *Avogadro's number*, of any substance represents a mole of that substance. A stoichiometric equation not only represents the nature of the substances entering the reaction; it also specifies the number of moles of each reactant. For example, we saw that two moles of hydrogen, added to one mole of oxygen yield two moles of water. Throughout this book every chemical symbol, unless otherwise specified, designates chemical concentrations in moles per liter of that substance.

2.3. RATES: DEFINITIONS

Let us consider a simple chemical reaction $A \to C$. If the concentration of the product C of the reaction at time t_1 is C_1 and at a later time t_2 is increased to C_2, we define the average rate or velocity of the reaction as

$$v = \frac{C_2 - C_1}{t_2 - t_1} = \frac{\Delta C}{\Delta t}$$

The instantaneous rate of the reaction is found, when t approaches the limiting value of zero.

$$v = \lim_{\Delta t \to 0} \frac{\Delta C}{\Delta t} = \frac{dC}{dt} \qquad (2.1)$$

A chemical rate can also be defined following, for example, the decrease in the concentration of reactant A. In this case we add a minus sign to the definition of the rate, $v = - dA/dt$, because the differential of a decreasing quantity is always negative; however, we must ensure that the rate of a chemical reaction remains positive definite. For the general reaction, Eq. (1.2), the rate is defined as $-(1/v)(dA/dt)$, where v is the stoichiometric coefficient of A in the reaction. In this way the absolute value of the rate is independent of the choice of the substances that enter its definition. (We know from experience that heating influences the rates of chemical reactions, so they also must be defined for a given temperature.)

2.4. EXPERIMENTAL MEASUREMENT OF RATES

The earliest experiment to determine the rate of a chemical reaction was performed by Wilhelmy in 1850. He measured the rate of transformation of sucrose into fructose and glucose in an acidic solution, showing that the rate at each instant was proportional to the amount of sucrose still present at that instant in the reaction medium.

The change in time of the concentration of reactants or products may be measured by several different methods. For example, one technique measures the electrical conductivity of ionic species, another the light absorption of colored substances. For some reactions, direct evaluations of chemical species is possible. With each method, the experimenter can ultimately trace a graph such as the one in Fig. 2.1. In this figure a function $A = A(t)$ is constructed

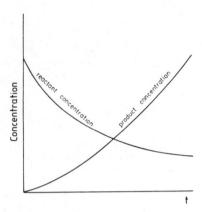

Fig. 2.1. Concentration of a reactant and a product versus time. Such curves may be constructed from experimental measurements.

experimentally for the concentration of one of the reacting chemicals. The slope of this curve at a given time is equal to $-(1/v)(dA/dt)$ (v is the stoichiometric coefficient), and therefore is the value of the instantaneous rate of the reaction as defined by Eq. (2.1).

Chemical rates themselves may vary in time. They are functions of the concentrations of the reactants and the products. The specific effect of concentrations on rates must be determined experimentally by the following procedure. For example, if chlorine molecules, Cl_2, are combined with nitrogen monoxide, NO, a new molecule is formed,

$$Cl_2 + 2NO \rightarrow 2NOCl \qquad (2.2)$$

The rate of this reaction can be measured easily under various conditions by methods similar to those outlined above. First, the concentration of NO may be held constant during the experiment. Now, if doubling the concentration of Cl_2 doubles the rate of formation of NOCl, it may be concluded that the reaction rate is proportional to the chlorine concentration. In another experiment Cl_2 is held constant and the relationship between the rate of NOCl formation and NO concentration is sought. Combining the two results, one finds

$$\frac{d[NOCl]}{dt} = k[Cl_2][NO]^2 \qquad (2.3)$$

which is called the *rate law* or *rate equation*.

Thus the rate or velocity of the formation of NOCl is proportional to the first power of Cl_2 and the second power of NO concentration, as two molecules of the latter enter into the reaction scheme. The proportionality factor k is called the specific *rate constant* or rate coefficient of the reaction. This coefficient may be estimated by the methods of the kinetic theory of gases, and is proportional to the number of collisions which produce new molecules. This is why the numerical value of the rate constant changes with temperature.

Equation (2.3) is a nonlinear (because of $[Cl_2][NO]^2 = xy^2$), first-order differential equation (because of $d[NOCl]/dt$), which gives the change in time of NOCl. Thus the study of reaction rates amounts to the solution of first-order differential equations, which often are non-linear.

2.5. REACTION ORDER

The term *reaction order* is used by chemists in another context, namely, the classification of reactions into well-defined categories. The reaction order is the exponent of a concentration factor that appears in the rate expression. In the example, Eq. (2.3), the reaction is of the first order with respect to chlorine concentration, Cl_2, and is second-order with respect to nitrogen monoxide, NO. The global order of the reaction is defined as the sum of the powers of all concentrations which appear in the differential equation of the rate law. We thus conclude that the reaction given in Eq. (2.2) is of the third order $(1 + 2 = 3)$.

Experimentally established expressions such as Eq. (2.3) may not always take a polynominal form. For some reactions, fractional or negative exponents may appear. For example, the rate of formation of hydrogen bromide in Eq. (1.3) is found to be

$$\frac{d[HBr]}{dt} = \frac{k[H_2][Br_2]^{1/2}}{1 + k'[HBr]/[Br_2]} \qquad (2.4)$$

and cannot be defined. Let us emphasize again that the order of a reaction has nothing to do with the stoichiometric coefficients and is entirely an experimental quantity.

2.6. ELEMENTARY REACTIONS

We have seen in Chapter (1) that a simple stoichiometric equation may hide a complex mechanism such as the one seen in Eq. (1.3). The latter evolves according to three *elementary steps* that follow each other in a well-defined sequence. An elementary step results from a single molecular encounter. Every elementary step has its own rate and participates in the establishment of the global rate of the reaction. A given elementary step may be intrinsically much slower than the other steps of the mechanism, and therefore essentially controls the global reaction rate. These slow reactions are called *rate determining steps*.

Elementary reactions are customarily classified according to the number of molecular species that enter the reaction step. This number defines the *molecularity* of the reaction. The elementary steps are unimolecular, bimolecular, and occasionally trimolecular. It is difficult to determine unambiguously the molecularity of an elementary step. The reactions reported in this book must be considered as illustrative examples.

2.6.1. Unimolecular Reactions

If a single molecule is transformed into one or more products by decay or rearrangement processes, the reaction is said to be unimolecular. For example, consider the following chemical transformation involving 1, 2 dichloroethylene molecules:

$$
\begin{array}{cc}
\mathrm{Cl} \diagdown \qquad \diagup \mathrm{Cl} & \mathrm{Cl} \diagdown \qquad \diagup \mathrm{H} \\
\quad \mathrm{C}{=}\mathrm{C} \rightleftharpoons \quad \mathrm{C}{=}\mathrm{C} \\
\mathrm{H} \diagup \qquad \diagdown \mathrm{H} & \mathrm{H} \diagup \qquad \diagdown \mathrm{Cl}
\end{array}
\tag{2.5}
$$

These two different molecules have the same atomic composition that we denote as $C_2Cl_2H_2$. However, the spatial arrangement of the atoms in the two molecular species is not the same. Such molecules are called *isomers*, and isomerization reactions are unimolecular.

If the rate of the above reaction is expressed in terms of the concentration of the reactant, 1, 2 dichloroethylene, experience shows that the rate decreases proportionally to the reactant concentration. Consequently

$$
-\frac{d[C_2Cl_2H_2]}{dt} = k[C_2Cl_2H_2]
\tag{2.6}
$$

In the right-hand side of this equation the product concentration appears to the first power; therefore all unimolecular reactions are of the first order. Moreover, the kinetic equation (2.6) reflects exactly the content of the stoichiometric equation (2.5).

2.6.2. Bimolecular Reactions

More often an elementary step is the result of the encounter of two atoms or molecules. For example, consider the reaction of chlorine with hydrogen molecules:

$$Cl + H_2 \rightarrow HCl + H \tag{2.7}$$

Experience shows that the rate of this reaction expressed in terms of chlorine molecules is given by

$$-\frac{dCl}{dt} = k[Cl][H_2] \tag{2.8}$$

Referring to the definition of reaction order, note that the rate given in Eq. (2.8) is of the second order $(1 + 1 = 2)$; therefore all bimolecular reactions are of the second order. Again, the rate equation (2.8) and the stoichiometric equation (2.7) give the same information about the mechanism of the reaction.

2.6.3. Trimolecular Reactions

If three molecules come together in a single encounter to form new chemical species, the elementary step is called *trimolecular*. An example of such a reaction is the encounter of two iodine atoms I and one inert species which produces molecular iodine I_2.

$$I + I + Ar \rightarrow Ar + I_2 \tag{2.9}$$

The role of the argon atom, Ar, is to absorb the excess energy that is released during the collisional process. The experimental rate of this reaction is

$$-\frac{dI}{dt} = k[I]^2[Ar] \tag{2.10}$$

and is therefore of the third order $(2 + 1 = 3)$. Trimolecular reactions are not very common, as the probability of triple encounters is much smaller than that of binary collisions.

*   *
*

Generally there is no relation between the order, molecularity, and stoichiometry of a reaction. The stoichiometric equation indicates only the overall number of moles and the nature of the substances participating in the reaction. Reaction order tells us about the mechanism of the reaction only in simple cases, whereas molecularity is only defined for each elementary step of a complex mechanism. The three different notions coincide only for the elementary steps. Whenever there is a discrepancy between the order and the stoichiometric equation, we must conclude that the reaction is complex and does not proceed via one simple, elementary step.

2.7. REVERSIBLE REACTIONS

In the preceding section we considered a few elementary reactions in which reactants transform into products completely and *irreversibly*. However, most often a portion of the newly formed product decomposes into the original reactants. If the reaction is left alone, it will reach an apparent quiescent state, with a well-defined concentration of reactants and products. For a given temperature, the overall yield of the reaction is the same as if we had started with pure products. We say that such a reaction is *reversible*. We shall illustrate with a simple example how to evaluate the rate of an elementary, reversible step.

Let us consider an *isomerization* reaction, the change in molecular geometry of fumaric acid into maleic acid known to proceed via a reversible mono-molecular step.

$$
\begin{array}{ccc}
\text{H—C—COOH} & & \text{H—C—COOH} \\
\| & \rightleftharpoons & \| \\
\text{HOOC—C—H} & & \text{H—C—COOH} \\
\text{Fumaric acid} & & \text{Maleic acid}
\end{array}
\qquad (2.11)
$$

Reaction (2.11) is composed of two monomolecular steps of the form: $A \overset{k_1}{\rightarrow} B$ and $B \overset{k_{-1}}{\rightarrow} A$, each with its own rate coefficient. A rate equation in terms of B can be established for both reactions. B is produced in the first reaction at a rate of $k_1 A$, and is decomposed in the second reaction at a rate of $k_{-1} B$; thus, the net rate of formation of B is the sum of the rates in the two elementary steps and reads

$$
\frac{dB}{dt} = k_1 A - k_{-1} B \qquad (2.12)
$$

We thus have found a general rule for establishing the rate of an elementary reversible step: global reaction rate = forward rate − reverse rate.

In a closed vessel with reactant A, the forward rate gradually decreases, and the reverse rate increases. At some instant they become equal, the global rate of the reaction reaches a zero value, and the reaction does not evolve anymore. The reaction has reached an *equilibrium state*.

If reaction $A \underset{k_{-1}}{\overset{k_1}{\rightleftharpoons}} B$ is at an equilibrium state then $k_1 A - k_{-1} B = 0$, and a new constant is defined as

$$\frac{B}{A} = \frac{k_1}{k_{-1}} = K \tag{2.13}$$

called the *equilibrium constant*. The equilibrium constant is very sensitive to temperature change, since k_1 and k_{-1} may depend on temperature in different ways. This is why heating of the reaction mixture usually favors one direction of a two-way chemical process more than the other.

By a similar line of reasoning, global rates and equilibrium constants can be defined for every chemical reaction. For the general scheme, Eq. (1.2), the equilibrium constant is given by

$$\frac{G^\gamma H^\delta \cdots}{A^\nu B^\mu C^\eta \cdots} = K \tag{2.14}$$

The equilibrium constant of a chemical reaction is a very important quantity, since it is a measure of the yield of the reaction. Large values of K indicate the formation of large quantities of new products. Equation (2.14) is one of the most important and widely used relations in chemistry, and is often referred to as the *mass action law*.

2.8. DETERMINATION OF RATE CONSTANTS

Specific rate constants, generally denoted by the small letter k, are deduced from experimental data. As an example, let us examine a well-known reaction, the decomposition of gaseous nitrogen pentoxide N_2O_5. The stoichiometric equation reads

$$2N_2O_5 \xrightarrow{\ k\ } 4NO_2 + O_2 \tag{2.15}$$

It is customary to write the value of the rate constant above the reaction arrow. Experimental measurements of the rate of this reaction show first-order kinetics

$$-\frac{dN_2O_5}{dt} = kN_2O_5 \tag{2.16}$$

despite the fact that two moles of N_2O_5 enter the stoichiometric equation (2.15). This example illustrates the fact that order and stoichiometry are two different notions.

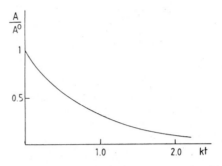

Fig. 2.2. If reactant A decomposes according to a first-order kinetics, its normalized concentration decays exponentially in time.

Equation (2.16) is a first-order linear differential equation that can be integrated immediately and yields

$$A(t) = A_0 e^{-kt} \tag{2.17}$$

Here the constant quantity A_0 is the initial amount, and $A(t)$ is the time-dependent concentration of N_2O_5. The time decay of nitrogen pentoxide is shown in Fig. 2.2.

Taking the logarithm of Eq. (2.17) yields

$$\ln A = \ln A_0 - kt \tag{2.18}$$

This equation has the functional form $y(t) = a - kt$ and represents a straight line with slope $-k$, as seen in Fig. 2.3. The numeric value of the slope k may be determined straightforwardly. Similar techniques can be used for determination of rate coefficients of higher-order reactions.

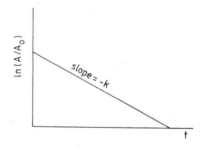

Fig. 2.3. Measurement of rate coefficients. The logarithmic plot of concentration against time is a straight line with slope $-k$.

2.9. DETERMINATION OF THE MECHANISMS OF CHEMICAL REACTIONS

We have already said that most chemical processes evolve by a succession of elementary steps. The ensemble of these steps constitutes the *reaction mechanism*. In principle, given a knowledge of the molecular species entering a reaction and their structure, one should be able to deduce the mechanism of the reaction from the laws of quantum mechanics. Unfortunately, as yet we are far from this goal, and must find indirect ways for the elucidation of reaction mechanisms. Usually one guesses a probable mechanism and checks its validity against experiments. Such guesswork relies on intuition, reasoning, the search for similarities between reactions in related families, and some clues from experimental measurements. Today chemists have a wealth of sophisticated physical and chemical methods for the detection of short-lived intermediate substances appearing in minute quantities (e.g., isotopic tracers, the detection of radicals, stereochemical evidence). The assumed mechanism must be consistent with the global stoichiometry of the reaction, the experimental rate expressions, and their temperature dependence.

As an example, consider the reaction between the hypochlorite OCl^- and the iodide ions I^-, in aqueous solution.

$$I^- + OCl^- \xrightarrow{k} OI^- + Cl^- \tag{2.19}$$

The experimental rate of this reaction expressed in terms of iodide ions is given by

$$-\frac{d[I^-]}{dt} = \frac{k[I^-][OCl^-]}{[OH^-]} \tag{2.20}$$

The presence of hydroxyl ions OH^-, absent in the stoichiometric equation (2.19), indicates that the reaction is not bimolecular but follows a complex mechanism. Various clues from different directions lead us to assume the following mechanism:

(1) $$OCl^- + H_2O \underset{k_{-1}}{\overset{k_1}{\rightleftharpoons}} HOCl + OH^-$$

(2) $$HOCl + I^- \xrightarrow{k_2} HOI + Cl^-$$

(3) $$HOI + OH^- \underset{k_{-3}}{\overset{k_3}{\rightleftharpoons}} OI^- + H_2O \tag{2.21}$$

Steps 1 and 3 are fast and reversible, while Step 2 is bimolecular and slow;

consequently, according to the "bottleneck" principle, the rate of the reaction is essentially controlled by the rate of this slow reaction. (This principle can be illustrated by the following example. The rate of evacuation of a crowded subway car is determined by the number of people who can cross the car's door per second. Therefore, the slow step is called the *rate-determining* step.)

From the scheme in Eq. (2.21) deduce the rate of iodide ions as determined by the slow Step 2

$$-\frac{d[I^-]}{dt} = k_2 [HOCl][I^-] \tag{2.22}$$

As Step 2 is slow, it may be assumed that the reversible Step 1 will have time to almost reach and remain close to its equilibrium state at all times. Following the same reasoning as for Eq. (2.13), from the first step of Eq. (2.21) we deduce.

$$k_1 [OCl^-][H_2O] - k_{-1}[HOCl][OH^-] = 0$$

As H_2O is the solvent, its concentration remains practically constant; therefore

$$k_1 [OCl^-] - k_{-1}[HOCl][OH^-] = 0 \tag{2.23}$$

Solve this equation for HOCl and introduce its value in Eq. (2.22). Recover the experimental rate equation (2.20), defining $k = k_2 k_1 / k_{-1}$.

$$-\frac{d[I^-]}{dt} = \frac{(k_2 k_1 / k_{-1})[OCl^-][I^-]}{[OH^-]} = \frac{k[OCl^-][I^-]}{[OH^-]} \tag{2.24}$$

Therefore, it is probable that the reaction of Eq. (2.19) follows a mechanism given by Eq. (2.21). When applying this type of method for the determination of reaction mechanisms, one must be aware that several different mechanisms may provide the same stoichiometric and rate equations. Elucidation of a complex mechanism is a difficult and tricky enterprise and may lead to diverging opinions among chemists.

2.10. MATHEMATICAL EVALUATION OF CONCENTRATIONS

It is important for chemists to be able to follow time changes of concentrations of different substances entering the reaction process. Of course, one can always try to measure the concentrations of different products in the course of time; but this procedure may become very tedious and time consuming. Moreover, some intermediary products may have very short lifetimes, and may not be detected easily.

Fortunately, if the mechanism of the reaction is known, there are mathematical methods that can enable one to follow the time-course of the reaction. This can be done simply with a pencil and a piece of paper. This procedure shall be illustrated with a simple model mechanism. Let us consider a two-step reaction,

$$A \xrightarrow{k_1} B$$

$$B \xrightarrow{k_2} C \tag{2.25}$$

Following the methods of the preceding sections, write the rate equations for all substances participating in the reaction as

$$\frac{dA}{dt} = -k_1 A$$

$$\frac{dB}{dt} = k_1 A - k_2 B$$

$$\frac{dC}{dt} = k_2 B \tag{2.26}$$

The time evolution of concentrations $A = A(t)$, $B = B(t)$, and $C = C(t)$ can be found by integration of this set of coupled first-order differential equations. As initially only substance A is present, with a given concentration A_0, we find

$$A = A_0 e^{-k_1 t}$$

$$B = \frac{k_1 A_0}{k_2 - k_1} (e^{-k_1 t} - e^{-k_2 t})$$

$$C = A_0 - A - B \tag{2.27}$$

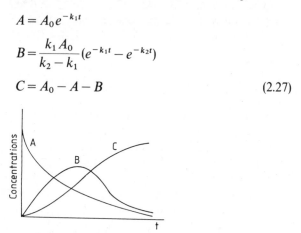

Fig. 2.4. Time variation of A, B, and C as computed from Eq. (2.27).

From Eq. (2.27) we can compute for any given time the amount of A, B, and C present in the reaction. This time variation is reported in Fig. 2.4. For more complex reactions, try to reduce the number of differential equations and mathematical difficulties encountered by using appropriate approximations. Whenever analytical calculations are impossible, integrate the rate equations by numerical methods.

2.11. CATALYSTS

The rate of a chemical reaction may increase dramatically by the addition of minute quantities of substances called *catalysts*. Catalysts do not appear in the equations and often are not even altered by the chemical process. Catalysts are used on a large scale in industry. For example, the octane rating of gasoline is increased if hydrocarbon-normal butane is transformed into isobutane.

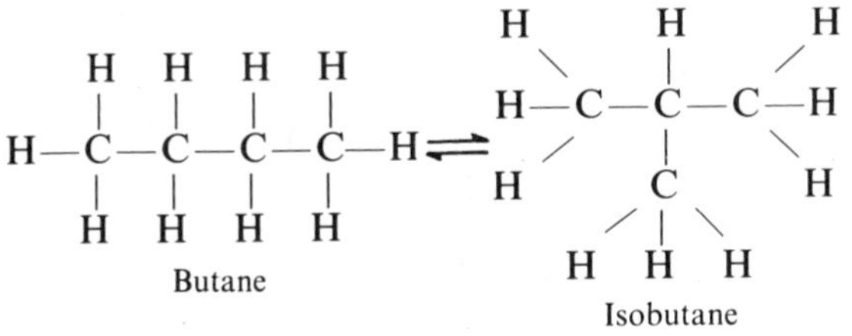

$$\tag{2.28}$$

Butane

Isobutane

This isomerization reaction is catalyzed by aluminum chloride.

The most spectacular catalysts are *enzymes* that function in living media. They catalyze many thousands of very complex and vital biochemical reactions in living organisms, which proceed at relatively low temperatures. These reactions would otherwise be impossible.

*   *

*

To appreciate the unexpected nature and various aspects of the self-organization of matter, which has been a recent discovery, it is necessary to go back to the science of thermodynamics. For two centuries the macroscopic laws of equilibrium thermodynamics predicted the gradual disorganization of matter toward a final disordered state. Only by extending thermodynamic methods to matter which is no longer in a quiescent state can we understand the underlying mechanisms of the self-organization of physical systems.

Chapter 3

EQUILIBRIUM THERMODYNAMICS

3.1. WHY THERMODYNAMICS?

In all their fundamental aspects, the cells that compose a living organism are physico-chemical systems. Their activity results from and produces chemical energy. The food we eat and the effort we produce are the consequence of transformations of this energy, which will be defined more precisely later. Our preoccupation with the number of calories contained in a given quantity of foodstuff reflects our concern with chemical energy. Chemical reactions which occur inside cells are called *biochemical processes*, and are to be studied by the same physico-chemical laws as ordinary chemical reactions.

In Chapter 2 we showed how to evaluate and measure the rate of a chemical reaction. Now it is necessary to answer such questions as: Will a given reaction occur spontaneously? Can the reaction accomplish biologically meaningful work? What is the yield of a given chemical reaction? How can one increase or decrease this yield? How much energy will be released or absorbed by a given biochemical reaction?

When the reactions are at equilibrium, such questions may, in principle, be answered in the framework of a macroscopic theory known as *equilibrium thermodynamics*, provided we can isolate part of the living cell and consider it as a chemical system in equilibrium. As we shall see later, not all the properties of living systems may be explained with such a procedure. Life being an eminently dynamic process, a modified version of equilibrium thermodynamic methods must be developed.

In what follows we shall investigate the properties of the simplest possible thermodynamic system, a uniform gas expanded or compressed in a cylinder with the help of a piston moving slowly enough to preserve the uniformity of the gas. After the basic concepts are introduced, we shall consider more complex situations, but without leaving the basic piston and cylinder and the uniform gaseous state of matter. This will constitute the *system*, that is, the portion of the real world we wish to investigate. The reason for this choice is twofold. First, only with such a simple system is it possible to present the basic concepts and laws of equilibrium thermodynamics in an elementary fashion. The second reason is of historical and practical nature; thermodynamics was developed as a means of improving systems which were essentially of the piston and cylinder type.

The theory we are about to develop is so general that it may be applied to man as well as to simple machines. Moreover, in the framework of this theory we are able to understand many experiences of our daily life. Why does a cup of hot coffee eventually cool off? Why does the smell of perfume fill an entire room? More importantly, why in the present era of supertechnology can't we find a way to extract cheap energy from the oceans or from the surrounding atmosphere?

The contribution of thermodynamic theory to the understanding of matter and the physical world may be divided, somewhat arbitrarily, into three categories:

1. Thermodynamic theory provides quantitative values for various properties of the system.
2. It gives information about the possibility and impossibility of processes and shows the direction of evolution of a macroscopic system.
3. It provides methods for testing the stability of a given state of the system.

Chapters 3 and 4 will be devoted to that part of thermodynamic formalism which furnishes quantitative values about the state of a given system. Chapter 5 considers the evolutionary aspect of thermodynamic theory, while stability properties are investigated in Chapter 6.

3.2. THE SCIENCE OF FIRE

Fire has fascinated and terrorized the human race throughout its history, but by the time of the great Ice Ages, humans had learned to tame fire into a constructive source of useful heat. Until the end of the eighteenth century, fire was mainly used for heating, cooking, melting, and as a source of light. In some civilizations of antiquity, fire was a subject of worship and an agent of purification. In ancient Persia fire symbolized Ahura Mazda, the god of Zoroastrianism. In Greek mythology, Prometheus saved the human race by bringing the celestial gift of fire from the sun. The Aztec, Norse, and Hindu pantheons also have their gods of fire. That fire also generates power can be seen by anyone watching a covered boiling kettle of water. Heron of Alexandria made use of hot vapor in the construction of the first eolipile (early gas turbine), which was used as a miracle producer in temples and a gadget of amusement for the wealthy.

The new industrial society of the seventeenth and eighteenth centuries needed coal at an ever increasing rate. Rising water in the deep coal mines had to be eliminated, and muscle power was too slow and inefficient to do this. In 1690 the Frenchman Papin conceived of the first vacuum-producing steam pump. A few years later such pumps were operational in English mines thanks to the ingenuity of Savery, Newcomen, and Fulton. In 1765, James Watt modified the extremely inefficient Newcomen pump into a more efficient device by using a thermodynamic property, the adiabatic expansion, and also by introducing an automatic control inside the engine. The pump was transformed into the first *steam engine*. A steamboat operating with a Watt engine made its first journey in 1807.

This revolutionary discovery opened up the era of heat engines or "machines." Engines drastically changed the nature of the human societies that developed them, turning them into what are known today as "industrial societies." The social upheaval that resulted was termed the *industrial revolution*. The development of heat engines was followed by a theoretical approach to the problem of the interrelationship between heat and mechanical motion. In 1824 Sadi Carnot became the founder of this new branch of macroscopic science, called thermodynamics.

It was soon realized that the same laws, properly modified, could be used for the study of other properties of bulk matter. For example, one can predict the effect of change of pressure, temperature, and composition on a great variety of physico-chemical systems and many important processes of life.

If we open a book on thermodynamics, we discover at most four laws and an impressive number of mathematical formulas. A close scrutiny reveals that only a few quantities enter into these endless mathematical transformations, that sometimes look very similar but in reality are different. This "dryness" in the presentation of thermodynamics makes it a somewhat unpopular and abstract subject among the other macroscopic sciences. However, thermodynamics is with us on a daily basis. With the help of thermodynamics, we can understood how our car functions, predict why water boils, why a lump of sugar dissolves faster in hot coffee than in cold, evaluate how much energy we gain by eating one lump of sugar, and many other daily events related to our immediate experience with nature, such as the formation of clouds, rain, or snow.

The power of thermodynamics lies in the fact that it describes and correlates directly observable properties of different substances. This is done in the absence of any detailed knowledge of the internal structure of the bulk matter. With relatively few laws and variables, an impressive number of remarkable conclusions can be drawn for complex systems containing a great number of individual molecules. Another advantage of thermodynamics is that it is model-independent. The specific nature of substances enters into the theory by a few parameters, such as heat capacities or molar volumes. The part of thermodynamic formalism which provides information about the behavior of a system may be divided into two distinct parts. The formalism of *equilibrium thermodynamics* is only concerned with the description of the state of a system here and now, and does not consider the history and the path by which a given state is reached. On the other hand, *nonequilibrium thermodynamics* is concerned with the dynamic aspect of bulk matter, and therefore with systems that are not in a state of rest.

The spontaneous self-organization of matter is only possible in non-equilibrium systems. However, nonequilibrium thermodynamics cannot be understood without some knowledge of the most salient features of equilibrium thermodynamics.

3.3. A THERMODYNAMIC SYSTEM: DEFINITIONS

A *thermodynamic system* is defined as that portion of the physical world of macroscopic size which we intend to investigate. Let us consider four typical thermodynamic systems as shown in Fig. 3.1. A piston *B* holds *n* moles of a gas of uniform composition inside a rigid cylinder *D*. The piston is thermally isolated from the gas. The whole constitutes a thermodynamic system that is referred to as a *BD* system. If the piston and cylinder are insulated and protected completely from all external influence, the thermodynamic system is *isolated* (Fig. 3.1.a). If the system is only prevented from exchanging heat and matter with its surroundings, we are dealing with an *adiabatic* process (Fig. 3.1.b). Without the insulating jacket, *BD* may exchange heat or other forms of energy (to be defined later) with its surroundings, and thus becomes a *closed system* (Fig. 3.1.c). Finally, an *open system* (Fig. 3.1.d) exchanges both matter and energy with its surroundings.

The *n* moles of the pure gas at temperature *T* inside the cylinder occupy a volume *V* and exert a force per unit of area *p*, called pressure, on the piston. These are called *state variables*; they define completely the *state* of the thermodynamic system here and now, independent of the system's previous history. However, as we shall see later, these variables may be related to each other. If one of these variables is modified, the system undergoes a *process* and reaches a new state.

Some variables of the system, such as temperature *T* and pressure *p*, are defined locally and are independent of the size of the system. These are *intensive variables*. On the other hand, if, for example, at constant pressure and temperature the number of moles of the gas in the cylinder changes, the volume *V* will also change. Such global properties of the system are described as *extensive* variables. If a given parameter is prescribed as a constant or as a given function of time, it becomes a *constraint*, whereas if it is left to evolve

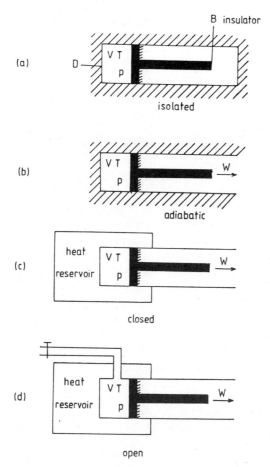

Fig. 3.1. A piston B holds a gas of uniform composition at pressure p, temperature T, and volume V inside the cylinder D. (*a*) An isolated system: there is no exchange of heat, work, and matter with the surroundings. (*b*) An adiabatic process: no heat or matter is exchanged with the surroundings. (*c*) A closed system: heat and work are exchanged with the surroundings. (*d*) An open system: heat, work, and matter are exchanged with the surroundings.

freely under the action of other constraints, it becomes a *variable*. As we shall see later, a given state of the system may be characterized by such *state variables* linked together by an *equation of state* of the form $f(V, T, p) = 0$. Indeed, intuition and experience tell us that the variables of a given system are not independent. For example, for a sufficiently diluted gas, the variables of the system are related by the simple equation of state.

$$pV = nRT \qquad (3.1)$$

R denotes a well-defined constant and T is the so-called absolute or thermodynamic temperature defined below. Equation (3.1) is the well-known law of *ideal gases*.

3.4. EQUILIBRIUM THERMODYNAMICS

Let us assume that the piston in Fig. 3.2 is clamped at a position S_1 in the cylinder, which contains a compressed gas at room temperature. Now if we suddenly release the piston, it will first oscillate but finally will come to rest at position S_2. At this position three antagonistic forces act on both sides of the piston: the atmospheric pressure and the weight of the piston push it inward, while the pressure of the gas inside the cylinder acts in an outward direction. The equality between these forces determines the equilibrium position S_2. It is in this sense that we speak of a thermodynamic system in a *state of mechanical equilibrium*. Before reaching position S_2 the system was in a *mechanical non-equilibrium state*. Quite generally, when thermodynamic processes reach final and permanent states of rest, and when the corresponding matter and energy fluxes vanish, we may speak of *thermodynamic equilibrium states*. For example, a cup of hot coffee, if left alone, reaches *thermal equilibrium* with its surroundings and we speak of coffee "at room temperature." When a chemical reaction does not evolve any further, it has reached a *chemical equilibrium* state. However, if because of constant constraints matter and energy fluxes do not vanish, then thermodynamic equilibrium is replaced by a *thermodynamic stationary* state.

The concept of equilibrium is central to all aspects of thermodynamic

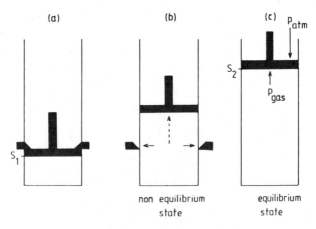

Fig. 3.2. A state of mechanical equilibrium. (*a*) The cylinder holds a compressed gas. (*b*) p_{gas} pushes the piston upwards. (*c*) An equilibrium state is reached.

formalism. As we shall see, when a system undergoes a finite amount of change, we may describe the phenomenon by a succession of a great number of quasi-equilibrium states, provided the rate of change is not too high. For example, in *BD* system the motion of the piston must be sufficiently slow to preserve the uniformity of the gas and to avoid the propagation of internal waves.

3.5. THE ZERO LAW AND THE CONCEPT OF TEMPERATURE

The scientific investigation of any natural phenomenon usually requires a means of quantitative evaluation of some properties of that phenomenon. Thus, the science of fire required a quantitative evaluation of the subjective sensations of "hotness" and "coldness." We saw that in Aristotelian philosophy, hot and cold were associated with terrestrial elements. Therefore, it was natural for Greek scientists, although not too keen on the experimental approach, to investigate the qualitative differences between these two elements. The first thermoscope was developed by Phylon of Bysantia in 250 B.C. This rudimentary device was able to distinguish between a balloon filled with cool air and the same balloon exposed to the heat of the sun. Heron of Alexandria later constructed a more refined thermoscope, but in the social conditions of the Hellenistic world and the Middle Ages that followed, these devices remained solely objects of amusement for more than 1500 years.

During the Renaissance, with the revival of science and arts in the newly formed bourgeois societies, the works of Heron of Alexandria became accessible to Italian scientists. Santorio, in 1630, modified the thermoscope of Heron in such a way as to make it suitable for measuring the body temperature of his patients. A few decades later, the first real thermometer was developed in Florence. In the next century, several varieties of thermometer were developed. The German Fahrenheit (1715) and the Swede Celsius (1742) invented the two most popular temperature scales. The invention of the thermometer not only made it possible to quantify the "hotness" of objects, but also, as we shall see later, made possible the measurement of the energy contained in a given mass of matter.

In the development of the first and second laws of thermodynamics, the concept of temperature was intuitively related and limited to the thermometer readings. But as the science of thermodynamics developed, a new law was needed that would introduce and define temperature as an intrinsic property of any thermodynamic system, independent of any specific thermometer scale. The zero law of thermodynamics fulfills this need. It is based on the following considerations.

We know from experience that a set of objects may be arranged in a unique sequence, according to their hotness. Water vapor is warmer than liquid water

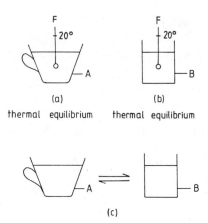

Fig. 3.3.   The zero law. F is in thermal equilibrium with A and also with B. A and B are in thermal equilibrium.

which in turn is warmer than ice. On the other hand, objects of different hotness if put in thermal contact will finally reach thermal equilibrium and exhibit the same temperature. The zero law is based on this property and may be formulated as follows: *If two objects* A *and* B *are independently in thermal equilibrium with a third object* F *then* A *and* B *are in thermal equilibrium with each other.* In Fig. 3.3 the spoon is in thermal equilibrium with the coffee in the cup. If we put the same spoon into a glass of water and if the temperature of the spoon does not change, the spoon is also in thermal equilibrium with the water. The zero law states that the coffee and water are also in thermal equilibrium and therefore are of the same temperature.

Ordinary mercury thermometers are constructed on this principle. When a thermometer is in thermal equilibrium with a mixture of crushed ice and water, its temperature is arbitrarily set to show a reading of zero degrees; if the thermometer is in equilibrium with boiling water at one atm., its reading is set at 100 degrees on the Celsius scale. Quite generally, the zero law may be used to define the concept of temperature independently of the nature of thermometric substance, such as mercury or alcohol. The numerical scale, however, may depend on the properties of standard substances.

3.6.   THE FIRST LAW OF THERMODYNAMICS

The first law of thermodynamics introduces the familiar and at the same time the extremely abstract concept of *energy*. Broadly speaking, this term describes the capacity for any animate or inanimate object to do work, that is, to make an effort. Prior to the development of heat engines, the only useful energy for the betterment of the human condition stemmed from muscular power and mechanical devices. However, the possibility of the conversion of heat into mechanical work was demonstrated successfully with the advent of steam engines.

The beginning of the nineteenth century was a fertile period in the experimental sciences. Various new phenomena were discovered, one after the other. Galvani created the first electrical circuit. Volta discovered that chemical reactions may produce electricity, and constructed the first electrical battery. Electric current could also initiate chemical reactions, produce heat, and be associated with magnetic phenomena. Faraday generated electric current by magnetic forces. Seebeck showed that heat may produce electrical current, and finally Pelletier in turn used electricity in order to decrease the temperature of objects. Therefore, it appeared that chemical reactions, heat, electricity, and mechanical devices all have the capacity to do work. Moreover, this capacity was convertible from one set of phenomena into the next. In the mid nineteenth century, Joule made a quantitative study of the conversion of mechanical work into heat and showed their equivalence.

All these observations lead to the conclusion that "something" conveniently termed *energy*, with the capacity to do work, may manifest itself under several disguises. These findings were generalized by R. Meyer and Helmholtz (1847) in what is known as the *law of conservation of energy*, which constitutes the essence of the first law of thermodynamics, which states that *all macroscopic systems have definite and precise amounts of energy. The quantity of energy of the system may only change if energy is added or subtracted from the system. No energy can be created or destroyed.*

It is difficult to answer in a simple way to the question. "What is energy?" We know of energy only through its manifestations when it generates change. Consequently, we can only measure differences of energy, and only these differences have any physical meaning. Nevertheless we shall continue to speak of energy, although what we really mean is differences of energy. Now we want to relate the abstract concept of energy to the known and tangible properties of thermodynamic systems.

In a macroscopic description of matter, every object is associated with a given quantity of energy. This energy may be subdivided into three contributions. For example, *potential energy*, ΔE_p, may be generated by the gravitational forces of our planet and is dependent on the position of the object. A weight of one pound lying on the ground is harmless, but it can have significant power if it falls from a skyscraper. An object also has *kinetic energy*, ΔE_k, which is the expression of the motion of the object. A little bullet becomes dangerous only when it is ejected with high speed from a gun barrel. Finally, an object is endowed with a given quantity of *internal energy*, ΔU_I, by its sheer existence. This energy is associated with the translation, rotation, and vibration of atoms, as well as with the energy of electrons and nuclei, and more generally it includes the energy associated with all structural forces inside matter. Therefore

$$\Delta E = \Delta E_k + \Delta E_p + \Delta U_I \tag{3.2}$$

Usually a thermodynamic system is investigated under conditions such that ΔE_p is irrelevant and ΔE_k vanishes; for example, the piston and cylinder stand firmly on a table.

In such cases $\Delta E = \Delta U_I$, and in thermodynamics, when we talk about energy, we mostly mean such internal energy differences. There are important engineering applications when ΔE_k does not vanish, but they will not concern us in this book.

Before we go any further, we must find a mathematical expression for the familiar mechanical energy called *work*, which is defined in the following manner. If a force $F = Mg$ acts on a mass M submitted to the gravitational force and lifts it some infinitesimal distance dx from a reference point, a given quantity of work is performed. We know intuitively that if M or dx is doubled, twice as much work is produced. Therefore, it is natural to write that the infinitesimal work dW is proportional to these two quantities:

$$dW = Mgdx = Fdx \tag{3.3}$$

The total amount of work performed in lifting the mass M from point Z_A to Z_B is thus given by

$$W = \int_{z_A}^{z_B} Fdx = Mg(z_B - z_A)$$

3.6.1. Measurement of Internal Energy

Let us consider an adiabatic, or thermally isolated, frictionless piston and cylinder full of gas, as shown in Fig. 3.4. All forces acting on the piston are such that the latter has reached and remains in equilibrium at position S_1.

If the external forces are changed by an infinitesimal amount, the gas will push the piston upward and reach a new equilibrium position S_2.

If the surface area of the piston is A and it has traveled an infinitesimal distance dl during the expansion process

$$dW = pA\,dl = p\,dV$$

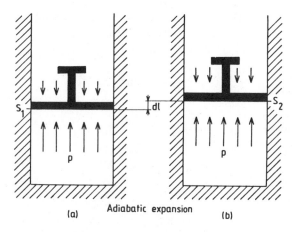

Fig. 3.4. (*a*) A gas in equilibrium state in an adiabatic BD system. (*b*) The piston has traveled a distance *dl*; *dW* work has been performed by the gas.

work has been performed by the gas. A finite amount of work may be obtained if the process of expansion, as described above, may be repeated many times. Therefore

$$W = \int_{V_1}^{V_2} p \, dV \tag{3.4}$$

In the process of expansion, the gas furnishes W work to the outside world. But as the system is adiabatic and no other forms of energy may enter into the cylinder, according to the law of conservation of energy, the gas has lost internal energy. Therefore,

$$\Delta U = -W \tag{3.5}$$

If we compress the piston and bring it into its initial position, by successive infinitesimal changes dl, W work is performed on the system and the lost internal energy is restored back to the gas.

The insulating jacket of the cylinder may be replaced by a thermostat (heat reservoir) and the expansion and compression processes are therefore performed isothermally. Again, if the expansion occurs slowly, in the presence of an ideal gas, it can be shown that $dW = dQ$. The transfer of dQ heat from the reservoir to the gas has increased the work capacity of the latter.

Quite generally, if a thermodynamic system receives Q quantity of heat and furnishes W work to the outside world, according to the first law the change in the internal energy of the system is given by

$$\Delta U = U_{\text{fin}} - U_{\text{ini}} = Q - W \qquad (3.6)$$

Here Q and W are both defined as positive quantities. If $\Delta U = 0$ the system after performing *a cycle* returns to its original state, $Q = W$, and the initial state variables are recovered.

3.6.2. Internal Energy Function

The fundamental equation (3.6) describes a simple bookkeeping procedure. In any mathematical formulation of the physical world, we need functions expressed in terms of relevant and accessible variables that describe a given system. However, the beauty of thermodynamics stems from the fact that assuming the existence of thermodynamic functions, one can elaborate a complete mathematical theory and obtain a large amount of useful information about relations between thermodynamic quantities. Therefore, we only need to postulate the existence of an unknown *internal energy function U*, and determine the variables that enter this function. However, we keep in mind that only ΔU has any physical meaning.

We are able to make several guesses about the energy function. For a given volume, we know that the more gas there is in the cylinder of Fig. 3.4, the more work it can perform. Therefore, energy is an extensive quantity. On the other hand, we saw that the energy of a system may be increased by the addition of heat. At constant volume, this will increase the temperature of the system. Finally, a gas at high pressure may perform more work than at low pressure. From these considerations we conclude that energy is some unknown function of temperature T, pressure p, and number of moles n, or $U = U(T, p, n)$.

In Fig. 3.4 the quantity of gas in the cylinder is constant; therefore, when the piston is at point S_1 the system is characterized by T_1, p_1, and the energy function is therefore $U_1(T_1, p_1)$. After expansion, the energy function $U_2(T_2, p_2)$ describes the system at S_2, and is only determined by the values of T_2, p_2 at this point, and is independent of the path followed by the system to reach S_2. Thus energy is a *state function*, and T_1, p_1, T_2, p_2 are *state variables*. Usually heat and work are not state functions, as a given change of ΔU can be achieved by many different combinations of Q and W,

$$\Delta U = Q_\alpha - W_\alpha = Q_\beta - W_\beta = \cdots = Q_\lambda - W_\lambda \qquad (3.7)$$

Let us express the first law of thermodynamics for an infinitesimal change of

dU; it reads

$$dU = dQ - dW = dQ - pdV \qquad (3.8)$$

and

$$\Delta U = \int_{S_1}^{S_2} dU = U_{S_2} - U_{S_1}$$

Thus, dU is an exact differential as it expresses the differences of energy between two states; but in general, we cannot make similar statements for dQ and dW, which are not exact differentials.

3.7. THE SECOND LAW OF THERMODYNAMICS

3.7.1. Reversible Processes

As we already said, the major role of a heat engine is to transform, with a maximum efficiency, thermal energy into mechanical motion. In the example of Fig. 3.4, after removing the insulating jacket, we could introduce a quantity of heat Q inside the cylinder and transform it into W work. The first law predicts the total convertibility of these two forms of energy inside the cylinder. However, because of frictional forces the motion of the piston in the cylinder generates heat, and a part of the mechanical energy is turned into thermal energy which is dissipated into the surroundings.

The loss may be minimized but not supressed if the piston is made as frictionless as possible. This is one of the reasons why all mechanical parts in an engine need lubrication. In what follows, for theoretical convenience we continue to assume the existence of a totally frictionless piston. But even an ideal and totally frictionless piston and cylinder does not entirely convert heat Q into work W. To see this, we add Q heat to the BD system and release the piston suddenly. The piston and the weight oscillate, eventually reach an equilibrium position, and a given quantity of work has been performed. However, a fraction of the available mechanical energy remains in the gas in the form of internal energy because of the viscous processes that followed the expansion of the gas. This internal energy is eventually dissipated into the surroundings and cannot be recaptured and converted into work by the expanding gas. This point will become clear later on.

Frictional and viscous forces are approximately proportional to the velocity of the expanding piston. Therefore, these forces may be minimized if the piston travels extremely slowly; the dissipation of heat into the surroundings is also minimized. At the limit of an infinitely slow motion of the piston, a maximum amount of work is produced by the expanding gas. Unlike the dissipated thermal energy, lost mechanical energy may be returned to the gas by a slow process of compression. Such processes are called *reversible* and may be defined as follows: *if the direction of a thermodynamic process can be reversed at any moment by an infinitesimal change in the surroundings; and, in addition, if the sign of heat exchange may be reversed, the process is reversible.*

We may wonder how much slowness is acceptable in a real world. For example, for the specific example of the *BD* system we can imagine that the piston, originally at equilibrium state of pressure p_1, has expanded by an amount dV to another equilibrium state of pressure $p_1 + dp$. We assume that the time necessary for the gas to reach the new state of equilibrium is of the order of $\tau = 10^{-3}$ seconds. Now if the time of the expansion of the gas is of the order of $\tau' = 10^{-1}$ seconds, the piston has traveled very slowly between the two states of equilibrium relative to the equilibration time.

For a long time, reversible processes were central to thermodynamic studies, as these were the only processes for which a theoretical approach was easy. During the expansion process, at each step the state of the system is only infinitesimally far from the equilibrium state. Consequently, following the concepts of Section 3.4, thermodynamic variables such as temperature and pressure may be defined for such reversible processes.

The concept of reversibility is not limited to mechanical processes; heat may also be transferred along a reversible process. This happens whenever the heat transfer is practically without influence on the uniformity of temperature required from an equilibrium system. It must always be remembered that reversible transformations are *ideal approximations* for the actual, near-equilibrium natural processes, which to some degree are all irreversible.

3.7.2. Irreversible Processes

Experience shows that matter has the tendency to diffuse spontaneously from more concentrated areas into less concentrated ones. For example, an open bottle of perfume will evaporate irreversibly into the surroundings; a cup of hot coffee will reach and remain at room temperature; a glass of whiskey on the rocks will warm up to the same room temperature. From these examples and other daily experiences, we must conclude that some natural processes follow, spontaneously and irreversibly, a privileged direction of change. In order to recover the initial state, we need to act upon the system by another process the natural direction of which is opposite to that of the original process.

In all the examples cited above, if we imagine a perfectly isolated room, the total energy of the room and the various containers is conserved. Nothing in the first law, which expresses the conservation of energy, prevents the heat from the surrounding air from entering the coffee cup, bringing it to the boiling point. The wandering perfume molecules may return into the bottle without violating the first law. Therefore, it is obvious that the zero and first laws do not suffice for a complete description of the thermodynamic behavior of a system. In order to be able to make predictions about the restrictions imposed on the direction of evolution of natural processes, we need a new law which introduces a new property.

In equilibrium thermodynamics, we only deal with state functions and are not interested in the history or the trajectory by which a state is reached. A cup of coffee is hot or it is cold. The details and the knowledge of the path (the process of cooling) between these two states are beyond the methods of equilibrium thermodynamics and the associated quasi-equilibrium processes cited above. Thus the new property must also introduce an additional state function characterized, at least near equilibrium, by the state variables, such as p, T, and n. On the other hand, this new property must predict the spontaneous and natural direction of change of a given process.

The zero law of thermodynamics was associated with the concept of temperature. The first law embodies the concept of conservation of energy. We must expect that a new property will introduce a new law in the formalism of equilibrium thermodynamics. This is indeed the case, and now we are ready to introduce the second law of thermodynamics.

3.7.3. The Birth of the Second Law

The natural direction of the evolution of processes may be harnessed to do useful work. In the example of the piston and cylinder of Fig. 3.4, we saw that a compressed gas has the natural tendency to expand and may lift a weight M, thus performing a given amount of mechanical work. This property formed the basic principle upon which the early pumps, the Watt heat engine, and, later on, the familiar automobiles were constructed.

In a heat engine, the piston and the cylinder are connected to other mechanical devices that transform the mechanical work of the moving piston into a rotary motion of the wheels of the engine. A substantial amount of work may only be obtained (some distance traveled by the relevent forces) if the motion of the piston is repeated periodically. This is only possible if the piston is forced to return to its initial state, and if a flux of heat is regularly supplied to the gas. The piston must perform work in a periodic motion (recovering the same position and velocity) while the *BD* system travels along a *thermodynamic cycle*; that is, the system recovers its initial state variables.

In 1824, because of the already considerable importance of steam engines, Sadi Carnot undertook a mathematical analysis of these engines and opened the way to a theoretical study of the science of fire. Carnot's aim was to determine whether there was a limit to the improvement of heat engines. He soon realized that the most efficient steam engine is the one that performs an ideal reversible cycle. The piston must travel continuously between a hot and a cold reservoir. Carnot found that in such conditions all conceivable cycles have the same maximum efficiency, which is a function of the temperature of the two reservoirs. In 1850, following Carnot's discovery, Kelvin stated the content of the *second law* in the following manner: "It is not possible to devise an engine which working in a cycle shall produce no effect other than the extraction of heat from a thermostat and the performance of an equal amount of mechanical work." Therefore, a Carnot engine cannot produce useful work if it is only hooked to a single reservoir. In other words, a heat engine working in a cycle will only produce work if it is inserted between at least two reservoirs of different temperature. If the second law did not exist, we could have resolved the recent energy crisis by extracting cheaply the gigantic amount of energy contained in the oceans and the atmosphere.

Relying on the second law, Kelvin deduced the so-called *absolute thermodynamic temperature scale* with the following properties: (1) it is independent of any particular substance, and (2) there is an absolute zero, the same for all substances. Henceforth T will denote the thermodynamic temperature scale.

Later on, Clausius gave a mathematical formulation to the findings of Carnot and Kelvin. His formalism is so general that it can be applied to any system. It can predict the direction of change of the hot coffee, the cold whiskey or the evolution of the universe. In Clausius's time, the principle of conservation of energy was already established, and could be incorporated in his formalism. Clausius assumed the existence, for every thermodynamic system, of an abstract state function *entropy* with the following properties. *The entropy of an isolated system increases until a maximum value is reached.* This law, and its consequences, are known as *the second law of thermodynamics*. Let us now derive the second law with the help of intuitive arguments.

Before we define a state function accounting for the directional evolution of thermodynamic systems, we must express the irreversibility in terms of the measurable properties of that system. We have already said that a given volume expansion will produce less work than the corresponding reversible process, if the transformation is carried out irreversibly. The irreversible process causes a smaller decrease in the internal energy of the gas than the reversible process, and the remaining energy increases the temperature of the system. If we take out of the hot gas, by a reversible transformation, Q quantity of heat, we may bring the temperature and pressure to the same value as those obtained for the reversible process, and the two states of the gas become completely identical (see Fig. (3.6)). Therefore, Q is a measure of the irreversibility of the thermodynamic process. But we have already seen in Section 3.6 that the change in the internal energy of a system brought about by the exchange of Q quantity of heat is path-dependent; and therefore Q is not a state function and cannot serve our purpose. However, we know from experience that whatever a given quantity of heat can do depends very much on the temperature at which it is delivered. There is more heat in a swimming pool than in the flame of a candle burning for five minutes; but the candle flame melts a piece of wax instantaneously, while the same wax only becomes wet if immersed in the swimming pool.

From these considerations we may guess that some combination of Q and the temperature T at which it is delivered can characterize the irreversible evolution of a system, provided it defines a path-independent state function. Indeed, with the help of thermodynamic cycles, which shall not be reproduced here, Clausius could show a quite general property, namely, that for all reversible and irreversible cycles one has

$$\oint_{\text{cycle}} \frac{dQ}{T} \leqslant 0 \begin{cases} < 0 \text{ irreversible cycle} \\ = 0 \text{ reversible cycle} \end{cases} \tag{3.9}$$

(Here $\oint$ denotes the integral over a cycle.) The infinitesimal heat exchange dQ is defined positive if it is received by the system, and is negative otherwise.

For a reversible cycle the path between two arbitrary points A and B is shown in Fig. 3.5, and we have

$$\int_{A\alpha B} \left(\frac{dQ}{T}\right)_{\text{rev}} + \int_{B\beta A} \left(\frac{dQ}{T}\right)_{\text{rev}} = 0$$

or

$$\int_{A\alpha B} \left(\frac{dQ}{T}\right)_{\text{rev}} = -\int_{B\beta A} \left(\frac{dQ}{T}\right)_{\text{rev}} = \int_{A}^{B} \left(\frac{dQ}{T}\right)_{\text{rev}}$$

This equality shows that the new property is independent of the arbitrary

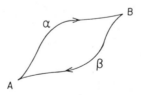

Fig. 3.5. Entropy change in a cyclic process. A reversible transformation brings the system from state A to state B along two distinct paths, α and β.

paths α and β and is only a function of the initial and final states A and B. Therefore, as for internal energy, it may be written as a difference between two values of a state function S

$$S_B - S_A = \int_A^B \left(\frac{dQ}{T}\right)_{rev} \tag{3.10}$$

The property may be generalized to irreversible transformations and reads

$$S_B - S_A > \int_A^B \left(\frac{dQ}{T}\right)_{rev} \tag{3.11}$$

These properties led Clausius to associate with the second law a new state function called *entropy S*, which shows the direction of evolution of physical and chemical processes.

For an infinitesimal change of entropy function along a reversible and irreversible path we have

$$dS = \left(\frac{dQ}{T}\right)_{rev} \qquad\qquad \text{reversible processes}$$

$$dS > \left(\frac{dQ}{T}\right)_{rev} \qquad \text{or} \quad dS = \left(\frac{dQ}{T}\right)_{rev} + d_i S \qquad \text{irreversible processes}$$

$$\tag{3.12}$$

$(dQ/T)_{rev}$ is the flow of entropy due to the interaction with the surroundings, whereas $d_i S$ is the part of entropy increase resulting from irreversible processes inside the system. We shall come back to this part of entropy in the next chapter.

For an isolated system there is no entropy flow. Therefore from Eq. (3.12) we deduce the fundamental inequality of equilibrium thermodynamics

$$d_i S \geqslant 0 \begin{cases} = 0 & \text{reversible processes} \\ > 0 & \text{irreversible processes} \end{cases} \tag{3.13}$$

For example, for an adiabatic process ($dQ = 0$) the entropy of the system increases ($d_i S > 0$) and reaches a maximum value ($d_i S = 0$) at equilibrium state.

3.7.4. The Entropy of an Ideal Gas

We shall now evaluate the increase in entropy of a simple system that undergoes a reversible process; and we shall verify that the entropy function as defined above has all the desired properties, namely, that entropy is a state function of well-defined state variables of the system, and moreover is an extensive quantity.

Let us go back to the piston and cylinder of Fig. 3.2 and fill it with n moles of an ideal gas. According to the first law, the infinitesimal change in the internal energy of such a system is given by $dU = dQ - dW$ (see Eq. (3.8)). The work dW results only from the expansion of the gas and therefore $dW = pdV$. On the other hand, the pressure of an ideal gas is given by the equation of state, Eq. (3.1), $p = nRT/V$. We conclude that $dW = nRTdV/V$. On the other hand, the heat capacity at constant volume of a substance is defined as $C_V = (dQ/dT)_V$. If the volume remains constant, from the first law we find $dQ = dU$. Thus, for a two-variable ideal gas (n fixed) $C_V = (\partial U/\partial T)_V$ and from the first law we deduce

$$dQ = dU + dW = C_V dT + \frac{nRT}{V}dV$$

From the definition of entropy, Eq. (3.9), we see that

$$dS = \frac{dQ}{T} = C_V \frac{dT}{T} + \frac{nR}{V}dV$$

From this expression we may conclude that entropy $S = S(V, T, n)$ is a function of state variables T, V, and n, and exhibits the following properties:

$$S_B - S_A = C_V \ln\left(\frac{T_B}{T_A}\right) + nR \ln\left(\frac{V_B}{V_A}\right) \tag{3.14}$$

From this expression we may conclude that entropy $S = S(V, T, n)$ is a function of state variables T, V, and n, and exhibits the following properties:

1. The entropy of a system is equal to the sum of the entropies of its different parts. In our example entropy increases with the number of moles of the gas enclosed in the cylinder, and therefore it is an extensive quantity.

2. Entropy is a state function. The entropy change between two states A and B is only a function of the state variables. In our example these are T_A, T_B, V_A, V_B, and n; thus entropy is independent of the various paths that may join these two states.

The equation of state of an ideal gas $pV = nRT$ may also be used to express

$V = V(p, T, n)$. The above calculations may be repeated with this function, and one shows easily that $S = S(p, T, n)$. Properties (1) and (2) of entropy are quite general and are not limited to the specific case of the ideal gas.

3.7.5. Direction of Change of Entropy

Let us show, with the help of a simple example, that the entropy increases in the presence of irreversible processes and remains constant if the processes are reversible.

To show this we isolate again the piston and the cylinder from their surroundings by an insulating jacket and assume that the gas in a state A inside the cylinder is at temperature T_1 and pressure p_1. We perform an adiabatic reversible expansion at a very slow pace and bring the pressure to a lower value p_2. The system now is in a state B and has produced some amount of work. No internal energy has been transformed into heat. As the system is isolated, no heat is exchanged with the surroundings, $dQ = 0$, and therefore the temperature also drops to a value T_2. Fig. 3.6 shows different steps of this reversible transformation. From Eq. (3.11) we deduce $\Delta S = S_B - S_A = 0$ and $S_B = S_A$. In this simple example, entropy is seen to remain constant during an adiabatic reversible process. By a reversible transformation, we can bring back the system to its initial state A. We start another expansion and bring the system to the same pressure p_2. However, this time we proceed rapidly, and therefore the process of change is irreversible. Part of the internal energy remains in the gas, and the system reaches a new state C at a higher temperature T_3. In this expansion the temperature and pressure have a

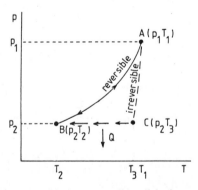

Fig. 3.6. Entropy remains constant during a reversible adiabatic transformation. It increases if the expansion is irreversible. By removing Q quantity of heat reversibly we may bring the temperature and the pressure to the same value as in B. (From H. C. Van Ness, 1975, reprinted by permission of author.)

meaning only in the initial and final states A and C. The broken line in Fig. 3.6 indicates that p and T do not have a unique value between A and C.

As a result of self-heating, we may suspect that entropy must have been produced inside the system. From Eq. (3.12) we see that we can only evaluate the quantity of entropy produced if we take the heat reversibly out of the cylinder. If we discard the insulating jacket, we could bring the gas from state C to state B and lower its temperature to T_2 by removing a given quantity of heat reversibly from the gas. As heat has been exchanged reversibly with the surroundings, from Eq. (3.10) we conclude

$$\int_C^B \left(\frac{dQ}{T}\right)_{rev} = S_B - S_C = \Delta S$$

However, as heat had been removed from the system, dQ is defined as a negative quantity. Consequently ΔS is also negative, which in turn implies that S_C is larger than S_B. But we know already that $S_B = S_A$, therefore we conclude that S_C is larger than S_A.

$$\int_A^C dS = S_C - S_A > 0 \tag{3.15}$$

This relation indicates an entropy increase during the irreversible expansion of the gas in the cylinder.

3.7.6. Why Does a Cup of Coffee Get Cold?

Let us reconsider the problem of a hot cup of coffee in a room. We assume that the room and the coffee cup together form the thermodynamic system under consideration, and that the room is perfectly isolated. The total isolated system undergoes irreversible processes; therefore, the entropy increases inside the room. As the entropy is an extensive quantity, from Eq. (3.13) we write

$$dS_r + dS_c = d_i S > 0 \tag{3.16}$$

Now dQ_r and dQ_c heat are exchanged between the room and the coffee; but as the system is isolated we have $dQ_r + dQ_c = 0$, which implies $dQ_r = -dQ_c$. Now if the room is at temperature T_r and the coffee at T_c the entropy production of the system may be deduced from Eq. (3.13)

$$d_i S = \frac{dQ_r}{T_r} + \frac{dQ_c}{T_c} = dQ_r \left(\frac{1}{T_r} - \frac{1}{T_c}\right) > 0 \tag{3.17}$$

If the coffee is warmer than the room, $(1/T_r - 1/T_c)$ is positive, therefore dQ_r must also be positive, and heat is transferred gradually from the coffee to the room until thermal equilibrium $T_c = T_r$ is reached between the two parts of the isolated system. In this case we deduce from Eq. (3.17) that $d_iS = 0$ in accordance with equilibrium properties.

3.8. THE THIRD LAW OF THERMODYNAMICS

There is a so-called third law of thermodynamics, also known as the theorem of Nernst, which, roughly speaking, may be stated in the following way: *at the absolute zero of temperature, the entropy of any system may be taken as equal to zero.* This law is of no consequence in this book, and therefore we shall not elaborate on it any further.

*   *

*

With the four laws and the relationships that were derived from them, the thermodynamic formalism is complete and may be applied to a wide variety of macroscopic systems. The exceedingly mathematical aspect of most books on thermodynamics stems from the fact that one must perform a lot of simple mathematical transformations in order to relate the various laws to measurable properties of real systems. A few examples are given in the next sections.

3.9. HOW TO USE THE THERMODYNAMIC FORMALISM

The beauty of equilibrium thermodynamics resides in the fact that with a few laws and the knowledge of relatively simple mathematics one may derive many general and useful relationships between various state variables of the system.

In the theory of equilibrium thermodynamics, if we introduce the definition $dS = dQ/T$ of entropy arising from the second law into the first law expressed by Eq. (3.8) we find:

$$dU = TdS - pdV \tag{3.18}$$

If we refer to the meaning of the total differential, we deduce that for the specific case of a gas of constant composition, the energy U is a function of the entropy and the volume of the system: $U = U(S, V)$. The total differential of such a function of two independent variables V, S is given formally by

$$dU = \left(\frac{\partial U}{\partial S}\right)_V dS + \left(\frac{\partial U}{\partial V}\right)_S dV \tag{3.19}$$

$(\partial U/\partial V)_S dV$ is the infinitesimal contribution to the internal energy when the volume changes by an infinitesimal quantity dV and the entropy remains constant. A similar interpretation is given to the second term of Eq. (3.19). Of course Eqs. (3.18) and (3.19) are equivalent, since dV and dS are arbitrary increments of independent variables. Therefore a direct comparison yields

$$\left(\frac{\partial U}{\partial S}\right)_V = T, \qquad \left(\frac{\partial U}{\partial V}\right)_S = -p \tag{3.20}$$

Furthermore, according to the rules of differentiation we see that

$$\frac{\partial}{\partial V}\left(\frac{\partial U}{\partial S}\right) = \frac{\partial}{\partial S}\left(\frac{\partial U}{\partial V}\right)$$

and with the help of Eq. (3.20) we find

$$\left(\frac{\partial T}{\partial V}\right)_S = -\left(\frac{\partial P}{\partial S}\right)_V \tag{3.21}$$

From this expression, known as *Maxwell's relation*, we see that we have interrelated all the relevant properties of the system. The measurements of some will give information about those quantities which are less accessible to direct experimentation. This endeavor is precisely the subject matter of the thermodynamics of the equilibrium states.

The internal energy function U is called a *thermodynamic potential*, since so many important quantities may be derived from it by differentiation.

3.10. THE GENERALIZED THERMODYNAMIC FORMALISM

The preceding sections introduced, with the help of the simple example of the piston and cylinder, the formalism of equilibrium thermodynamics. However, in general a thermodynamic system is much more complex. It is not necessarily isolated or closed, or of uniform composition, or without chemical reactions. Furthermore, the entropy S and the internal energy U are not the most convenient state functions for a direct evaluation of practical thermodynamic quantities. Temperature, pressure, and volume are more suitable variables, since they can be measured by common laboratory instruments. Moreover, in some instances, for example, in processes which evolve in rigid vessels, volume remains constant and pressure becomes the relevant state variable. Thus the above formalism must be modified to encompass much more general situations.

The approach to this problem does not introduce new laws or new concepts. It merely consists of defining new state functions which are combinations of U, S, and the state variables T, V, and p. From these new state functions, it is possible to deduce new Maxwell relations which are expressed in state functions other than S and U. A new state variable which may account for changes in the composition of the thermodynamic system is also introduced. Thus the formalism becomes suitable for the study of reaction mixtures and open thermodynamic systems.

3.10.1. Composition as a New Variable

In the example of the piston and cylinder a fixed quantity of a pure gas had been enclosed, within the cylinder, once and for all. Therefore composition was not a state variable, and we wrote simply $U = U(S, V)$. Let us enclose again in the cylinder a fixed amount of gas composed of hydrogen and nitrogen molecules. As we shall see later, these molecules may react together and produce a third gas, ammonia NH_3. Although the piston, the cylinder, and the gas form a closed system, the state of the gas is a function of the number of moles of the various substances enclosed in the cylinder. Thus the composition becomes a new state variable, and $U = U(S, V, n_{H_2}, n_{N_2}, n_{NH_3})$. We must keep in mind that, in a closed system, in the presence of chemical reactions all n_i are not independent variables. The total differential of this function is given by

$$dU = \left(\frac{\partial U}{\partial S}\right)_{V,n_i} dS + \left(\frac{\partial U}{\partial V}\right)_{S,n_i} dV + \sum_i \left(\frac{\partial U}{\partial n_i}\right)_{n_i',S,V} dn_i \qquad (3.22)$$

Here n_i' designates all the other composition variables but i. $(\partial U/\partial n_{NH_3}) dn_{NH_3}$ is defined as the contribution to the internal energy when dn_{NH_3} moles of ammonia are produced in the cylinder, while the entropy S, the volume V, and the number of moles of hydrogen and nitrogen are kept constant.

The third contribution in Eq. (3.22) is the change of internal energy generated by "chemical work." Naturally the chemical energy is proportional to the change in the number of moles dn_{NH_3}, dn_{H_2}, and dn_{N_2}. The coefficient of proportionality in Eq. (3.22) must necessarily be an intensive quantity similar to T and p. It is customary to designate this coefficient by μ_{n_i} and call it the *chemical potential.* Thus

$$\left(\frac{\partial U}{\partial n_{H_2}}\right)_{n'_i,S,V} = \mu_{H_2}, \qquad \left(\frac{\partial U}{\partial n_{NH_3}}\right)_{n'_i,S,V} = \mu_{NH_3}, \qquad \left(\frac{\partial U}{\partial n_{N_2}}\right)_{n'_i,S,V} = \mu_{N_2} \quad (3.23)$$

are, respectively, the chemical potentials of hydrogen, ammonia, and nitrogen in the cylinder.

In 1875 Gibbs introduced the concept of chemical potentials into the thermodynamic formalism and defined it in the following manner: "if, to any homogeneous mass, we suppose an infinitesimal quantity of any substance to be added, the mass remaining homogeneous and its entropy and volume remaining unchanged, the increase of the energy of the mass divided by the quantity of the substance added is the potential of that substance in the mass considered." Combining Eqs. (3.20), (3.22), and (3.23) we obtain the famous *Gibbs equation*:

$$dU = TdS - pdV + \sum_i \mu_i dn_i \qquad (3.24)$$

3.10.2. Open Systems

So far we have only been concerned with closed systems. However, open systems, characterized by an inflow and outflow of matter, are extremely important. With the introduction of the concept of chemical potentials, all thermodynamic relationships derived thus far may be extended to open systems in which all chemical variables are independent.

To see this, let us go back to the cylinder filled with a pure gas and inject a small quantity, dn, of the same substance, in the same thermodynamic state, into the cylinder. The gas in the cylinder is now an open system. As the internal energy of this open system is an extensive quantity, we expect that an additional term proportional to dn must be added to the energy equation (3.18), and from the definition of chemical potentials it follows that we must write

$$dU = TdS - pdV + \mu dn \qquad (3.25)$$

By comparing Eqs. (3.24) and (3.25) we conclude that the Gibbs equation is the most general expression which describes the change in the internal energy of a system when composition variables change either because of an external input or via chemical reactions. However, in a closed reacting medium all composition variables are not independent.

In order to become more familiar with the concept of chemical potentials, let us consider the following example. If we add to a bucket of water a drop of blue ink, the ink will gradually diffuse from high concentration regions into the low concentration zones of the bucket, and finally the water will take on a homogeneous pale-blue coloring. It so happens that the value of the chemical potential is higher in a concentrated ink solution than in a more dilute one. Therefore, chemical potentials perform the same function in matter flow as temperature gradients in heat flow. A difference in the chemical potential of a substance predicts its direction of flow. Matter always flows from regions of high into regions of low chemical potential. The physical nature of chemical potentials will become clearer when we treat the problem of the diffusion of coffee or perfume in an isolated room.

3.10.3. Other Thermodynamic Potentials

The entire formalism of equilibrium thermodynamics is derived from the fundamental Gibbs equation (3.24). However, volume, composition, and entropy are not the only relevant variables of a thermodynamic system. Moreover, the Maxwell relations, Eq. (3.21), are of limited interest in most experiments, since it is not easy to keep entropy constant. In some thermodynamic systems the volume remains constant and the pressure or temperature are the relevant state variables.

Thermodynamic formalism may be extended to such systems if one defines new state functions which are formed from combination of the state functions U, S and state variables p, T and V. One defines the *enthalpy* $H = U + pV$, the *Gibbs free energy* $G = U - TS + pV$, and the *Helmholtz free energy* $F = U - TS$. All these functions are thermodynamic potentials, and many important properties may be derived from them, as shown in Fig. 3.7.

Fig. 3.7 also shows that at equilibrium state the thermodynamic potentials G and F have minimum values, and therefore $dG = 0$ and $dF = 0$.

3.10.4. Physical Interpretation of Thermodynamic Potentials

If the pressure and composition of a system remain constant, $dn_i = 0$ and $dp = 0$, then from Fig. 3.7 we have $dH = dQ$. The change in the value of H for a process that brings the system from a state A to a state B is given by

$$\Delta H = H_B - H_A = Q_{AB}$$

Therefore, the variation at constant pressure of the state function enthalpy of a closed system is equal to the heat received by the system.

On the other hand, if the temperature and the composition of a one-

Thermodynamic Potentials

Internal energy	U	$U(S, V, n_i)$
Enthalpy	$H = U + pV$	$H(S, p, n_i)$
Helmhoelz free energy	$F = U - TS$	$F(T, V, n_i)$
Gibbs free energy	$G = U + pV - TS$	$G(T, p, n_i)$
	$= H - TS$	

Total Differentials of Thermodynamic Potentials

$$dU = TdS - pdV + \sum_i \mu_i dn_i$$

$$dH = TdS + Vdp + \sum_i \mu_i dn_i$$

$$dF = -SdT - pdV + \sum_i \mu_i dn_i$$

$$dG = -SdT + Vdp + \sum_i \mu_i dn_i$$

Some Identities

$$\left(\frac{\partial U}{\partial S}\right)_{V,n_k} = T \qquad -\left(\frac{\partial U}{\partial V}\right)_{S,n_k} = p \qquad \left(\frac{\partial U}{\partial n_k}\right)_{S,V,n_k} = \mu_k$$

$$\left(\frac{\partial H}{\partial S}\right)_{p,n_k} = T \qquad \left(\frac{\partial H}{\partial p}\right)_{S,n_k} = V \qquad \left(\frac{\partial H}{\partial n_k}\right)_{S,p,n_k} = \mu_k$$

$$-\left(\frac{\partial F}{\partial T}\right)_{V,n_k} = S \qquad -\left(\frac{\partial F}{\partial V}\right)_{T,n_k} = p \qquad \left(\frac{\partial F}{\partial n_k}\right)_{T,V,n_k} = \mu_k$$

$$-\left(\frac{\partial G}{\partial T}\right)_{p,n_k} = S \qquad \left(\frac{\partial G}{\partial p}\right)_{T,n_k} = V \qquad \left(\frac{\partial G}{\partial n_k}\right)_{T,p,n_k} = \mu_k$$

Fig. 3.7. Some definitions and identities.

component system are held constant, Fig. 3.7 shows that $dF = -pdV$; therefore,

$$\Delta F = \int_A^B dF = \int_A^B -pdV = -p(V_B - V_A)$$

This quantity represents the amount of useful expansion work performed by the system when the state of the latter is changed from A to B. From the definition $F = U - TS$ we see that the Helmholtz free energy is that part of the internal energy U which is not dissipated into waste energy or *unavailable energy* $T dS$.

At constant temperature and composition the change in the Gibbs free energy reduces to

$$\Delta G = \int_A^B dG = \int_A^B V dp$$

An explicit expression for ΔG may only be found if we can connect p to V. The equation of state of the ideal gas $V = nRT/p$ furnishes such a desired relationship and we find easily that

$$\Delta G = \int_A^B \frac{nRT}{p} dp = nRT(\ln p_B - \ln p_A) \tag{3.26}$$

It is customary to measure the free energy relative to a standard free energy g^0 taken as reference point. The latter is defined as the Gibbs free energy of one mole of the gas at the pressure of one atmosphere, and therefore is only a function of temperature T. Then if $p_A = 1$ designates the unit pressure in the same units as p, from Eq. (3.26) we get

$$G(T, p, n) = ng^\circ(T) + nRT \ln p$$
$$= n(g^\circ(T) + RT \ln p) = ng(T, p) \tag{3.27}$$

The function $g(T, p)$ is the Gibbs free energy *per mole* for an ideal gas, referred to a standard free energy level. From Eq. (3.26) we verify that the Gibbs free energy is a state function and is an extensive quantity. Such properties may be shown for all thermodynamic potentials of Fig. 3.7, since they all are defined as a combination of extensive state functions.

3.10.5. Molar Quantities

Molar quantities may be defined for all extensive properties of a thermodynamic system. For a system containing n moles of a gas we define *molar entropy*, *molar volume*, *molar enthalpy*, and *molar free energy* as:

$$s = \frac{S}{n}, \qquad v = \frac{V}{n}, \qquad h = \frac{H}{n}, \qquad f = \frac{F}{n}, \qquad g = \frac{G}{n} \tag{3.28}$$

It is obvious that all the thermodynamic relations derived so far are also valid for molar quantities.

Let us make a digression and introduce local thermodynamic quantities which are extensively used in the next three chapters. If N_γ is the mass fraction of γ component in a small element of an equilibrium system, v the specific volume, s the entropy, and u the energy per unit mass, then the fundamental equation (3.24) reads

$$du = Tds - pdv + \sum_i \mu_i dN_i \qquad (3.29)$$

The importance of this relation resides in the fact that it can be applied to any small volume element of a macroscopic system. It is also more general, as it may be extended to any point of a nonuniform medium. As we shall see in Chapter 4, when applied to a system undergoing irreversible processes, Eq. (3.29) represents the concept of *local equilibrium*.

Figure 3.7 shows that for a gas of uniform composition $\mu = (\partial G/\partial n)_{T,p}$; therefore from Eq. (3.27) we find immediately: $\mu = (\partial/\partial n)\,(ng(T,p)) = g(T,p)$. Thus we may conclude that the chemical potential of a substance is the increase in the Gibbs free energy of the system when one mole of that component is added to an infinitely large quantity of the same substance. Equation (3.27) provides an explicit expression for the chemical potential μ of a perfect gas

$$\mu = g^\circ(T) + RT \ln p \qquad (3.30)$$

Although this equation is derived for a pure and ideal gas, it may be generalized for the description of ideal gas mixtures and is currently used for the study of dilute solutions of various constituents in a solvent such as water. For example, if the concentrations of H_2, N_2, and NH_3 are sufficiently low in the cylinder and if p_{NH_3} is the partial pressure of the ammonia in the gaseous mixture, then

$$\mu_{NH_3}(T,p) = g^\circ_{NH_3}(T) + RT \ln p_{NH_3} \qquad (3.31)$$

Here the partial pressure of an ideal gas p_γ is related to the concentration C_γ by the equation of state $p_\gamma = n_\gamma RT/V = C_\gamma RT$. If we introduce this relation into Eq. (3.31) we find

$$\mu_\gamma(T, C_\gamma) = g^\circ_\gamma(T) + RT \ln RT + RT \ln C_\gamma$$
$$= \mu^\circ_\gamma(T) + RT \ln C_\gamma \qquad (3.32)$$

$\mu^\circ_\gamma(T)$ designates that part of the chemical potential which is a function only of the temperature of the ideal gas. Equation (3.32) may also describe the chemical potential of a constituent γ in a dilute solution, or a mixture of nonperfect gas. However, in this case $\mu^\circ(T, p)$ is a function of the pressure of the system.

Chemical potentials, always designated by μ_y are functions of composition variables. Throughout this book various units enter their definitions. The reader will find their meaning from the context.

3.11. CHEMICAL THERMODYNAMICS

As we have seen, in the eighteenth century it was realized that chemical energy is transferred in the course of reaction processes. As the thermodynamics of heat engines developed, it was soon discovered that the same formalism may also be applied to chemical reactions. In the nineteenth century the heat of reaction was chosen by Berthelot as a measure of the tendency for a chemical reaction to proceed. However, it turned out that this interpretation is only valid at $0°K$. As we already said Gibbs developed chemical thermodynamics and showed that the law of mass action (2.14) is an inevitable consequence of the laws of thermodynamics.

We have already seen that Gibbs introduced the concept of chemical potentials and showed how a change in the composition of a thermodynamic system may affect its various thermodynamic potentials. With the Gibbs equation (3.24), thermodynamics was ready to deal with chemical problems. We shall illustrate the main concepts of chemical thermodynamics with a simple example of a reaction proceeding entirely in the gaseous state.

3.11.1. Affinity of a Chemical Reaction

Ammonia, NH_3, is an active agent in some cleaning products and has many industrial and agricultural uses. It is synthesized by the combination of hydrogen and nitrogen molecules according to the stoichiometric relation $N_2 + 3H_2 \rightleftharpoons 2NH_3$.

This reaction was thoroughly investigated by Haber and served as a prototype for the elucidation of many of laws of gaseous-state chemical thermodynamics. During a reaction process evolving in a closed system, if dn_{NH_3} moles of ammonia are produced, then dn_{N_2} moles of nitrogen and dn_{H_2} moles of hydrogen have been consumed. These three quantities are not independent, but are related together by the stoichiometric equation that characterizes the reaction. Therefore, a single parameter must suffice to describe the progress of the reaction; and this can be found from

$$\frac{dn_{N_2}}{-v_{N_2}} = \frac{dn_{H_2}}{-v_{H_2}} = \frac{dn_{NH_3}}{v_{NH_3}} = d\zeta \qquad (3.33)$$

introduced in 1920 by De Donder and defines the *extent of reaction* ζ. $d\zeta$ may be understood as the amount of matter in moles that is exchanged from one side of the reaction to the other. The v's are the various stoichiometric coefficients as defined in Eq. (1.2). If v is positive, the reaction has moved from left to right, if negative it has evolved from right to left. For the synthesis of ammonia $v_{H_2} = -3$, $v_{N_2} = -1$ and $v_{NH_3} = 2$.

As nitrogen and hydrogen molecules disappear, ammonia is produced; thus all chemical potentials change during the reaction process. From Fig. (3.7) it is obvious that all four thermodynamic potentials, U, H, G, and F are also affected by this change. Moreover, if the reacting mixture is held at constant pressure and temperature, the Gibbs free energy $G = U - TS + pV$ is the most suitable thermodynamic potential for the investigation of chemical reactions. If dG_{Tp} designates the infinitesimal change at constant T and p

$$dG_{Tp} = \sum_i \mu_i dn_i = \mu_{NH_3} dn_{NH_3} + \mu_{N_2} dn_{N_2} + \mu_{H_2} dn_{H_2} \tag{3.34}$$

With the help of Eq. (3.33) we can eliminate all three dn_i from Eq. (3.34) and find

$$dG_{Tp} = (-v_{H_2}\mu_{H_2} - v_{N_2}\mu_{N_2} + v_{NH_3}\mu_{NH_3})d\zeta$$
$$= (\sum_\gamma \mu_\gamma v_\gamma)d\zeta = -A\,d\zeta \tag{3.35}$$

$A = -\sum_\gamma v_\gamma \mu_\gamma$ is defined as the *affinity* of the chemical reaction.

In Section 3.10 we saw that at thermodynamic equilibrium, $dG = -A d\zeta = 0$, which implies that necessarily the affinity A must be zero. (It is meaningless to imagine $A \neq 0$ and $d\zeta = 0$, as $d\zeta$ is an arbitrary infinitesimal increment.) Therefore,

$$v_{H_2}\mu_{H_2} + v_{N_2}\mu_{N_2} = v_{NH_3}\mu_{NH_3} \tag{3.36}$$

is the condition for the chemical reaction to be at equilibrium. However, before the equilibrium state is reached, the reacting mixture undergoes irreversible processes which produce NH_3 molecules; therefore, from Eq. (3.35) we have $dG = -A d\zeta < 0$. On the other hand, as $d\zeta$ indicates the degree of progress of the reaction, it is natural to think that the change in time of ζ, $d\zeta/dt = v$ defines the rate of the reaction. Therefore, the change in time of the Gibbs free energy reads

$$\frac{dG}{dt} = -A\frac{d\zeta}{dt} = -Av < 0 \tag{3.37}$$

and is only possible if A and v always have the same sign. For the chemical reaction under discussion, since ammonia molecules are produced continuously, the reaction velocity v must be positive; and therefore from Eq. (3.37) we conclude that the affinity A is necessarily positive:

$$A = v_{H_2}\mu_{H_2} + v_{N_2}\mu_{N_2} - v_{NH_3}\mu_{NH_3} > 0 \qquad (3.38)$$

This relation states that new ammonia molecules may be synthesized from hydrogen and nitrogen only if the chemical potentials of the latter multiplied by their respective stoichiometric coefficients exceed the chemical potential of ammonia times the corresponding stoichiometric coefficient. As the reaction proceeds, matter is transferred from higher to lower chemical potentials. Finally, the contribution to the affinity A from the chemical potentials of nintrogen and hydrogen balances the contribution from the newly formed ammonia molecules, and chemical equilibrium is established in the reaction vessel. In the equilibrium state, at the macroscopic level no synthesis or degradation of products is seen. Looked at from this angle, the affinity of a reaction and the underlying concept of chemical potentials play somewhat the same role as the temperature gradient in heat flow and the concentration gradient in matter flow.

Starting from the chemical equilibrium state, if we increase the amount of ammonia in the vessel at constant volume, the pressure of the gas increases, and the chemical potential of NH_3 also increases as $\mu_{NH_3}(T, p) = \mu^\circ(T) + RT \ln p_{NH_3}$. Thus the affinity for the synthesis of ammonia becomes negative, but at the same time the affinity for the production of hydrogen and nitrogen from the decomposition of ammonia is positive, and the reaction evolves in the opposite direction.

We owe to the Belgian De Donder the thermodynamic definition of affinity as described above. However, the word *affinity*, derived from the Latin *affinitas*, was first used by the alchemist Albertus Magnus in 1250 in the context of the transmutation of metals. The concept expressed the tendency of specific substances to react together and form new chemical species. The idea was revived in seventeenth century in the writings of Boyle, Newton, and Glauber. Already in 1718, the first "table of affinities" was established by Geoffroy.

3.11.2. Affinity and Progress of a Chemical Reaction

The concept of the equilibrium constant of a chemical reaction was introduced in Section 2.7. We have seen that this constant is a measure of the relative quantities of products and reactants in a given chemical reaction. The equilibrium constant K may be related to the various thermodynamic quantities in the following manner. We have seen that at chemical equilibrium the affinity of the reaction is zero; therefore, $\Sigma v_y \mu_y = 0$. Now if we introduce the explicit value of the various chemical potentials from Eq. (3.32) into the expression of A, we find that at equilibrium state

$$RT(v_{NH_3} \ln C_{NH_3} - v_{H_2} \ln C_{H_2} - v_{N_2} \ln C_{N_2})$$
$$= - v_{NH_3}\mu^\circ_{NH_3} + v_{H_2}\mu^\circ_{H_2} + v_{N_2}\mu^\circ_{N_2} = A^\circ \qquad (3.39)$$

For a gaseous mixture A° is a function only of the temperature of the system. Remembering the algebraic rules governing the logarithmic function, relation (3.39) may be written as

$$\frac{(NH_3)^{v_{NH_3}}}{N_2^{v_{N_2}}, H_2^{v_{H_2}}} = \exp\left(\frac{A^\circ}{RT}\right) = K \qquad (3.40)$$

The second equality follows from the definition of the equilibrium constant as seen in Eq. (2.14). $A^\circ = - \Sigma v_\gamma \mu^\circ_\gamma$ is also called the *standard affinity* of the reaction at the given temperature T. If we refer to the definition of μ°_i, the standard affinity may also be called *standard change of the free energy* of the chemical reaction.

From the definition of exponential functions we see that if A° is large then the equilibrium constant K is large; therefore, NH_3 is large and the reaction progresses in the direction of synthesis of ammonia molecules.

For a fixed temperature, $T = 25°C$, and $p = 1$ atm, values of standard affinities for common substances may be found in published tables. Therefore, for a given stoichiometric equation the value of the equilibrium constant K may be determined. This in turn fixes the ratio of products and reactants for the chemical process in its final equilibrium state, and gives a clue as to the direction of progress of the chemical reaction. At equilibrium state, if some quantity of NH_3 is added from outside into the reacting mixture, NH_3 decomposes into H_2 and N_2 molecules so as to reestablish the fixed ratio determined by Eq. (3.40).

A relation similar to Eq. (3.40) may be established for any arbitrary chemical reaction. For the general stoichiometric equation (1.2), one finds

$$\frac{G^\gamma H^\delta \cdots}{A^v B^\mu C^\eta} = \exp\left(\frac{A^\circ}{RT}\right) \qquad (3.41)$$

If the standard affinity A° is defined properly this equation may be used for solid, liquid and gaseous reacting mixtures.

We thus may conclude that if the stoichiometric equation of a chemical reaction is known, with the help of a table of standard affinities we may compute the equilibrium constant and therefore predict the possibility or impossibility of a given reaction without actually performing the experiment.

However, let us note that only the knowledge of the reaction rates may give us a clue about the possibility of self-organization in a reacting medium.

3.11.3. Why Does the Smell of Perfume Diffuse?

It is a common situation that if a bottle of perfume is opened, the smell soon fills the entire room. To understand the problem, we shall consider two completely isolated and separate rooms (see Fig. 3.8), held at a fixed temperature and pressure of 1 atmosphere. In Room 1 we open a bottle of perfume for 1 hour, while in Room 2 an identical bottle is left open only for 30 seconds. It is obvious that a heavy smell of perfume persists in Room 1, whereas the same smell is barely noticeable in Room 2. From the definition of chemical potentials we infer that $\mu_f^1 > \mu_f^2$. We now open the door between the two rooms and inquire about the distribution of perfume in the isolated system formed from the two.

We assume that the quantity of perfume is much smaller than the total amount of air; therefore, the air pressure in both rooms is practically equal and constant. Thus, from Eq. (3.30) we have $\mu_a^1 = \mu_a^2$. As T and p are constant, from Fig. 3.7 we see that the Gibbs free energy is the most appropriate thermodynamic potential for our problem, and as the latter is an extensive quantity we have

$$
\begin{aligned}
dG_{Tp} &= dG_{Tp}^1 + dG_{Tp}^2 \\
&= \mu_f^1 dn_f^1 + \mu_a^1 dn_a^1 + \mu_f^2 dn_f^2 + \mu_a^2 dn_a^2 \leqslant 0
\end{aligned}
\tag{3.42}
$$

However, the total system is isolated, and therefore the molecules of perfume and air must remain in the system; thus $dn_a^1 + dn_a^2 = 0$, and $dn_f^1 + dn_f^2 = 0$. These relations, together with the constancy of the chemical potential of the

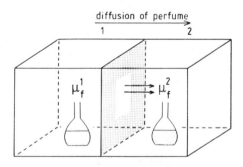

Fig. 3.8. A two-compartment isolated system. The direction of diffusional flow is determined by the condition $\mu_f^1 > \mu_f^2$.

air, reduces Eq. (3.42) to

$$dG_{Tp} = (\mu_f^1 - \mu_f^2) \, dn_f^1 \leqslant 0 \qquad (3.43)$$

As μ_f^1 is greater than μ_f^2, the only way to satisfy Eq. (3.43) is that dn_f^1 takes negative values. This in turn implies that perfume molecules must diffuse from Room 1 into Room 2.

3.11.4. Equilibrium Constant and Heat of Reaction

In Chapter 2, we saw that when two molecules of reactant combine together to form a product, the quantity of energy contained in various chemical bonds, called the chemical energy, is modified. Charcoal in the presence of oxygen burns and gives off heat. In chemical language this process is written as

$$C + O_2 \longrightarrow heat + CO_2$$

which must be completed by the conditions of the combustion process. The chemical energy stored in the chemical bonds of the C and O_2 molecules is greater than that stored in the bonds of the carbon dioxide, CO_2. As the total energy of the system must be conserved during the reaction process, the loss in the chemical energy appears in the form of heat and is called *heat of reaction*. Reactions which produce heat are called *exothermic*.

In another example, carbon may be combined with hydrogen to produce a hydrocarbon gas called ethylene

$$heat + 2C + 2H_2 \longrightarrow C_2H_4$$

Ethylene cannot be synthesized unless some quantity of heat, or *heat of formation*, is added to the mixture. Such processes are called *endothermic* reactions.

When a lump of suger, or glucose, $C_6H_{12}O_6$, is oxidized or "burned," it produces water and carbon dioxide according to the reaction

$$C_6H_{12}O_6 + 6O_2 \longrightarrow 6CO_2 + 6H_2O + heat$$

When we speak about the number of calories which are contained in a sugar lump, we precisely refer to this amount of heat. If we consume the same sugar lump and inhale oxygen, it will eventually transform into CO_2 and H_2O via a long succession of biochemical steps and produce the same amount of heat as if it had been burned outside the organism. In the living cell one part of the heat of reaction enters as heat of formation into various chemical bonds and the remaining part may be given off as heat by the organism.

In Section 3.10 we have shown that at constant pressure the heat exchanged between the system and its surroundings is equal to the corresponding change in the enthalpy of the system. It can be shown that in chemical systems subject to constant pressure and temperature the heat of reaction is equal to the difference of enthalpies of reactants and products. However, if h_i° is the standard molar enthalpy of i, then $\Delta H^\circ = \sum_i v_i h_i^\circ$, and

$$\text{heat of reaction} = -\Delta H^\circ$$

$$\text{heat of formation} = \Delta H^\circ$$

The negative sign in the definition of the heat of reaction follows from the fact that energy is lost by the reacting mixture, whereas the heat of formation is positive as enthalpy is added to the system.

With still more work, the heat of reaction or the heat of formation may be related to the equilibrium constant $K(T)$ by the van't Hoff equation

$$\frac{\partial \ln K}{\partial(1/T)} = -\frac{\Delta H^\circ}{R} \tag{3.44}$$

which, if integrated, furnishes the temperature dependence of the equilibrium constant K for a given ΔH°. Figure 3.9 shows such a variation. If we remember that the magnitude of K is a measure of the amount of reaction product, we see that as the temperature increases the quantity of product diminishes in an exothermic process, whereas in an endothermic reaction an increase in temperature enhances the yield.

From these considerations, we conclude that knowledge of the sign of the

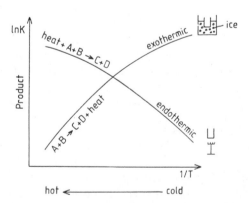

Fig. 3.9. Equilibrium constant K is a function of temperature. In an endothermic reaction, heat increases product formation. The yield of an exothermic reaction is increased by cooling.

heat of reaction makes it possible to act upon the equilibrium constant of the reaction and thus modify its yield.

3.12. THE LIMITS OF EQUILIBRIUM THERMODYNAMICS

We have sketched, in extremely broad lines, the theory of equilibrium thermodynamics and its importance for the study of chemical and biochemical processes. For the sake of simplicity, we have only considered perfect gas mixtures. However, the concepts and the various relations which have been obtained are quite general, and if modified slightly become still more powerful and may be applied to a great variety of systems, such as nonperfect gas mixtures, dilute solutions, solid mixtures, systems formed from several states of matter (multiphase systems), and mixtures undergoing several and simultaneous chemical reactions. Moreover, if electrical energy is also considered, the thermodynamic methods may be extended to electrically charged matter. In short, thermodynamic methods are a daily tool for engineers, physicists, chemists, and biologists, and are also used in other fields of scientific endeavor.

Although the formalism of equilibrium thermodynamics is a powerful tool for the study of physico-chemical systems, its use is limited by the fact that not every system is in thermal, mechanical, or chemical equilibrium. Many natural phenomena operate under the action of nonequilibrated forces, which in turn produce flows of various kinds and are therefore in nonequilibrium states. The majority of events in a biological system belong to this category of processes. Therefore, it is desirable to extend the macroscopic methods of thermodynamics to nonequilibrium systems which undergo irreversible processes. Chapter 4 is devoted to such an endeavor.

Chapter 4

NONEQUILIBRIUM THERMODYNAMICS

4.1. INTRODUCTION

In daily life we are faced with numerous irreversible phenomena. A cup of hot coffee left alone cools off and reaches room temperature, when it is in thermal equilibrium with the surroundings; or the cool air rushes in from an open window. Our very existence as a biological species is based on innumerable irreversible processes. Therefore, the understanding, prediction, and measurement of irreversible processes are central to any macroscopic view of natural phenomena. From the preceding chapter we infer that such an investigation might become possible from the knowledge of any change in entropy function, since entropy portrays irreversibility expressed in terms of meaningful and measurable properties of the thermodynamic system.

We have seen that in the construction of thermodynamic theory the concept of entropy followed immediately on the concept of energy. However, for a long time, the change in the entropy of a system could not be related to the tangible and measurable properties of irreversible processes. Therefore, the thermodynamic formalism was limited to the study of equilibrium states and reversible processes which were conveniently defined as a succession of quasi-equilibrium states. We have seen that in the equilibrium state, entropy is maximum and is a function of state variables U, V, and n. There is no need to know how entropy has reached its maximum value, and the only change in entropy results from an exchange d_eS with the surroundings during a reversible process. This exchange is given by the relation

$$d_eS = \left(\frac{dQ}{T}\right)_{rev}$$

(4.1)

as was shown in Eq. (3.12).

Another reason for neglecting entropy-producing processes was that for a long time they were considered essentially as a nuisance. They degraded useful energy into heat and prevented heat engines from producing a maximum yield as predicted by idealized reversible processes. Therefore, the only way to deal with them was to ignore or minimize their influence as much as possible. And thus, for a long time, thermodynamics was mainly concerned with equilibrium systems and reversible processes.

However, there were isolated attempts to treat irreversible processes by thermodynamic methods. As early as 1854 Lord Kelvin analyzed various phenomena with the help of thermodynamic methods. In the beginning of the present century Duhem, Jaumann, Lohr, and later Eckart attempted to relate irreversibility, expressed in terms of rate of change of local entropy, to the nonuniformity of a nonequilibrium system. In the same period De Donder derived a relationship which connected entropy change to the affinity of various chemical reactions.

In this chapter we shall follow a similar program, relating the irreversibility of processes, expressed as change in time of the entropy function, to the macroscopic and measurable properties encountered in evolving systems.

4.2. IRREVERSIBLE PROCESSES

A system in an equilibrium state is "dead"; it ignores time and history. This state is reached from a nonequilibrium and "alive" system endowed with "forces" that produce change, recognize the arrow of time, and drive the system gradually to its final death. This ineluctable fate may be avoided if, with the help of external forces, the system is kept artificially far from thermodynamic equilibrium. In nonequilibrium thermodynamics the terminology of force or *generalized forces* designates all agents of change, including the familiar mechanical forces.

For example, a constant input and output of energy and matter keeps a system in a permanent state of nonequilibrium. Irreversible processes are only seen in nonequilibrium systems. Therefore, in the context of the present discussion irreversibility and nonequilibrium are two faces of the same reality. The forces in a nonequilibrium system may, for example, arise because of the presence of chemical reactions or temperature or concentration differences in various parts of the system. Such a difference in a given quantity is called a *gradient* of that quantity. Forces produce currents, or *fluxes*, which in general tend to decrease the magnitude of the forces which generate them. All differences eventually disappear, and the system reaches the final state of equilibrium. For example, the onset of a driving force in the form of a temperature difference between two points of an object is followed by a transfer of Q quantity of heat per unit time and unit area from the hot to the cold region. This heat flux increases the temperature in the cold region at the expense of the hot areas and gradually drives the object to thermal equilibrium.

4.3. ENTROPY OF A NONEQUILIBRIUM SYSTEM

The presence of forces and fluxes in nonequilibrium systems implies that these systems are nonuniform or undergo chemical processes. Therefore, for the most general cases the composition and all the pertinent variables of the system may be time- and space-dependent. Thus, we may wonder how intensive quantities, such as temperature or pressure, can be defined for nonequilibrium systems; for example, the concept of temperature has been defined by the explicit use of the zero law of Section 3.5. However, these difficulties may be overcome if local equilibrium is assumed.

Let us consider a container full of coffee and kept over a weak source of heat. Far from the boiling point, a temperature gradient is established in the fluid. We divide mentally the total volume of the coffee into infinitesimal volume elements as seen in Fig. 4.1. We assume that the formalism of equilibrium thermodynamics can be applied to the local quantities of the small volume element. However, these quantities have different values for each volume element.

The local variables in Eq. (3.29) were defined per unit mass. However, it is more convenient to define new variables which describe properties in a unit volume. Then if ρ_γ is the mass per unit volume of γ component, and $\rho = \sum \rho_\gamma$, then $\tilde{s} = \rho s$ and $\tilde{u} = \rho u$ are respectively the entropy and energy per unit volume. Moreover as $\rho v = 1$, volume is no longer an independent variable.

We saw in Section 3.10 that the hypothesis of local equilibrium implies that

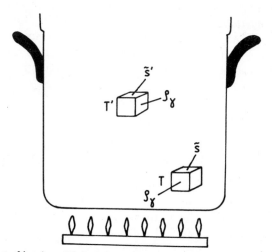

Fig. 4.1. Concept of local equilibrium. At each point of the nonequilibrium system (for each infinitesimal volume element) thermodynamic potentials are the same function of state variables as in the equilibrium state.

the local parameters are the same as in equilibrium states, and are linked by the same equation of state, that is, $\tilde{s} = \tilde{s}(\tilde{u}, \rho_1, \ldots \rho_\gamma)$. In other words, in every small volume the local entropy $\tilde{s}$ is related to the local energy $\tilde{u}$, and compositions ρ_γ by the Gibbs equation

$$T d\tilde{s} = d\tilde{u} - \sum_\gamma \mu_\gamma d\rho_\gamma \tag{4.2}$$

The total entropy of the system depicted in Fig. 4.1 is given by

$$S(t) = \int_V dV \tilde{s}(\tilde{u}, \rho_1, \ldots \rho_\gamma) \tag{4.3}$$

From this relation we deduce that the evaluation of the entropy of the coffee necessitates the knowledge of the local entropy function $\tilde{s} = \tilde{s}(\tilde{u}, \rho_1, \ldots \rho_\gamma)$.

We have just used a common trick of theoretical physics: when a given property of a system is not uniform in the entire volume, we try to find a local expression or a local evolution law for a local function and integrate the latter over the entire volume of the system. We shall make use of this procedure throughout this chapter.

4.4. ENTROPY PRODUCTION

As already stated, the irreversibility of a nonequilibrium system is expressed by the *entropy production*, which is a measure of the change in time of the entropy function. From Eq. (4.3), we conclude that we must somehow relate the change in time of the local entropy to physical quantities, such as generalized forces (gradients, affinities), and fluxes which actually characterize the irreversible processes inside the system.

Following the standard rules of differentiation of functions of several variables, we write

$$\frac{\partial \tilde{s}}{\partial t} = \sum_\gamma \left(\frac{\partial \tilde{s}}{\partial \rho_\gamma} \right)_{\tilde{u}, \rho_\gamma'} \frac{\partial \rho_\gamma}{\partial t} + \left(\frac{\partial \tilde{s}}{\partial \tilde{u}} \right)_{\rho_1 \ldots \rho_\gamma} \frac{\partial \tilde{u}}{\partial t} \tag{4.4}$$

Here ρ_γ are all independent variables and are no longer subject to the condition $\sum_\gamma \rho_\gamma = $ constant, which characterizes a closed system. From the local Gibbs equation (4.2) we find immediately that

$$\left(\frac{\partial \tilde{s}}{\partial \rho_\gamma} \right)_{\tilde{u}, \rho_\gamma'} = -\frac{\mu_\gamma}{T}, \quad \left(\frac{\partial \tilde{s}}{\partial \tilde{u}} \right)_{\rho_1 \ldots \rho_\gamma} = \frac{1}{T} \tag{4.5}$$

In the following sections we show that $\partial \rho_\gamma/\partial t$ and $\partial \tilde{u}/\partial t$ are given by the local laws of the conservation of mass of the individual components γ and of energy.

4.4.1. Change in Time of a Conserved Quantity

As we have seen, an irreversible process is essentially a dynamic one. Properties of the system change constantly in space and in time. We cannot quantify such restless and mobile systems unless we find laws which govern fluid flow. We were able to set the foundations of equilibrium thermodynamics with the help of the law of the conservation of energy. Let us look for other conserved quantities from our daily life. Intuition and experience tell us that the total mass of a closed system is conserved. Provided we neglect relativistic effects, mass cannot be created or destroyed. The amount of coffee in a cup diminishes only by drinking or evaporation. The quantity of the fluid only increases if we add fresh coffee.

A local formulation of conservation laws implies the use of a mathematical language that we would like to avoid. In this section we shall only give the final results and send the interested reader to more advanced books.

We consider a cube with six unit surface areas, as seen in Fig. 4.2. Flux $\mathbf{J}_d^\gamma$ is defined as the net amount of matter that crosses a unit surface area of the fluid per unit time in a given direction. The faster the fluid moves, the more matter will flow out of the cube. Moreover, a moment's reflection shows that the flux $\mathbf{J}_d^\gamma$ is proportional to the local density of matter, and is therefore given by $\mathbf{J}_d^\gamma = \rho_\gamma \mathbf{v}$. The total flux $\mathbf{J}_d^\gamma$ out of the cube thus is a directional quantity and in a Cartesian coordinate system takes the form

$$\mathbf{J}_d^\gamma = \rho_\gamma v_x \mathbf{1}_x + \rho_\gamma v_y \mathbf{1}_y + \rho_\gamma v_z \mathbf{1}_z = J_x^\gamma \mathbf{1}_x + J_y^\gamma \mathbf{1}_y + J_z^\gamma \mathbf{1}_z.$$

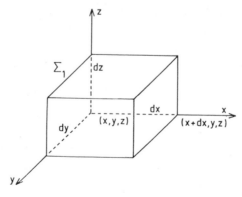

Fig. 4.2. Definition of the divergence of a vector $\mathbf{J}$. If $(\partial J_x/\partial x)\,dx\,dy\,dz$ is the flux across the two parallel surfaces Σ, the total infinitesimal flux is $\delta \mathbf{J} = (\partial J_x/\partial x + \partial J_y/\partial y + \partial J_z/\partial z)dx\,dy\,dz$.

As there are three orthogonal directions, one shows easily that

$$\frac{\partial \rho_\gamma}{\partial t} = -\left(\frac{\partial J_x^\gamma}{\partial x} + \frac{\partial J_x^\gamma}{\partial y} + \frac{\partial J_z^\gamma}{\partial z}\right) = -\operatorname{div} \mathbf{J}_d^\gamma \tag{4.6}$$

In this case div $\mathbf{J}_d^\gamma$ is a convenient shorthand notation for the expression on the right-hand side and serves as a definition for the mathematical object called *divergence of a directed quantity* $\mathbf{J}_d^\gamma$. The relation, Eq. (4.6), is precisely one of the missing terms in Eq. (4.4) for the local entropy production of an irreversible process.

If chemical reactions proceed inside the volume v, for every chemical species similar conservation laws may be written. Presently, we must distinguish between a contribution which results from the exchange of matter with the environment and an internal change of mass due to chemical reactions. Therefore, the total change in mass of a chemical species ρ_y is given by a balance equation made of the negative divergence of the flow of γ and a source term giving the production (or destruction) of substances. If r chemical reactions are present, we have

$$\frac{\partial \rho_\gamma}{\partial t} = -\operatorname{div} \mathbf{J}_d^\gamma + \sum_r v_{\gamma r} v_r \tag{4.7}$$

Here v_r is the rate and $v_{\gamma r}$ the stoichiometric coefficients of γ in the r reaction.

4.4.2. Change in Time of Local Internal Energy

The physical concepts and the mathematical formalism which we used in the derivation of Eq. (4.6) are not restricted to the fluid density. The same type of reasoning can be applied to the local internal energy function $\tilde{u}$, which is another conserved quantity. One finds that the change in time of the internal energy $\tilde{u}$, in the absence of convection processes, mechanical work, or any external forces, is given by

$$\frac{\partial \tilde{u}}{\partial t} = -\operatorname{div} \mathbf{J}_e \tag{4.8}$$

$\mathbf{J}_e$ is the *energy flux* and is defined as the quantity of energy that crosses a unit surface area per unit time in a given direction. If heat is the only form of internal energy that crosses the boundaries of the cube with a flux of $\mathbf{J}_q$, then Eq. (4.8) reduces to

$$\frac{\partial \tilde{u}}{\partial t} = -\operatorname{div} \mathbf{J}_q \tag{4.9}$$

4.4.3. Change in Time of Local Entropy

Introducing Eqs. (4.5), (4.6), and (4.8) in Eq. (4.4), after some rearrangement we find that

$$\frac{\partial \tilde{s}}{\partial t} = -\operatorname{div} \mathbf{J}_s + \sigma \qquad (4.10)$$

Thus contrary to the conservative quantities ρ_γ and $\tilde{u}$, the local change in time of the entropy function splits into two types of contributions. The first contribution is the divergence of a directional property and is therefore a flux term. This entropy flux $\mathbf{J}_s$ is caused by two different phenomena. One part is generated by heat flow $\mathbf{J}_q/T$, and another by diffusional fluxes of all chemical species given by $-(1/T)\sum_{\gamma=1}^{n}\mu_\gamma \mathbf{J}_d^\gamma$. The second contribution is a source or production term and is a quadratic form of all fluxes $\mathbf{J}_i$ and their corresponding generalized forces X_i. Therefore, if σ is the expression of *local entropy production*,

$$\sigma = \sum_i J_i X_i \qquad (4.11)$$

The vectorial notation has been dropped since in the general case all flows and forces are not always directional quantities.

Let us remember that the various inhomogeneities in gradients and chemical reactions which create fluxes are called generalized forces in order to distinguish them from purely mechanical forces. Fig. 4.3 shows the generalized forces and their corresponding fluxes which enter Eq. (4.11).

The local entropy production σ expresses the creation of entropy, in the presence of irreversible processes, inside the small volume element. From Fig. 4.3 and Eq. (4.11) we see that σ is formulated in terms of measurable and tangible thermodynamic quantities. Remember that all $\mathbf{J}_i$ and X_i vanish at thermodynamic equilibrium.

Our final aim is to evaluate $\partial S/\partial t$, the change in time of the total entropy S of the system. From the concept of local equilibrium of Section 4.3 we deduce that the total change of entropy is made of the sum of all local changes, and therefore

$$\frac{\partial S}{\partial t} = \int_V dV \frac{\partial \tilde{s}}{\partial t} = \int_V dV(-\operatorname{div} \mathbf{J}_s + \sigma) \qquad (4.12)$$

The second equality follows from Eq. (4.10). With the help of appropriate theorems, the integral of the entropy flux in the volume $V, \int dV \operatorname{div} \mathbf{J}_s$, may be transformed into a total flux of entropy $\partial_e S/\partial t$ across the surface

Irreversible Processes	Flows	Forces
Diffusion	$\mathbf{J}_d$	$\text{grad}\left(-\dfrac{\mu}{T}\right)$
Heat transfer	$\mathbf{J}_q$	$\text{grad}\dfrac{1}{T}$
Chemical reaction	v_i	$\dfrac{A_i}{T}$

Fig. 4.3. Generalized flows and their corresponding forces.

surrounding the volume V. We put $\partial_i S/\partial t = \int_V dV\sigma$ and write Eq. (4.12) as

$$\frac{\partial S}{\partial t} = \frac{\partial_e S}{\partial t} + \frac{\partial_i S}{\partial t} \tag{4.13}$$

Therefore, the total entropy in a given system changes only if entropy flows across the boundaries or is produced inside the system. There are no forces and fluxes in an equilibrium system; therefore, $\partial_i S/\partial t = 0$, and consequently $\partial S/\partial t = \partial_e S/\partial t$. We have just confirmed a well-established property of equilibrium systems, namely, that the only way entropy may change is for it to cross the boundaries of the system.

If we isolate the system from external entropy flux, from Eq. (4.13) we find

$$\frac{\partial S}{\partial t} = \frac{\partial_i S}{\partial t} > 0 \tag{4.14}$$

This inequality results from our knowledge that the entropy of an isolated system may only increase, thus $d_i S \geqslant 0$ (see Eq. (3.13)). Equation (4.14) is equivalent to the expression $\int_V dV(\sum_i X_i J_i) \geqslant 0$. The assumption of local equilibrium tells us that if a relation is valid for the total volume, it must also be valid for any small volume element of that system. Thus we go back to the local description and write

$$\sigma = \sum_i X_i J_i \geqslant 0 \tag{4.15}$$

Equation (4.15) states that the sum total of all $X_i J_i$ must be positive, which leaves open the possibility for any individual $X_i J_i$ to become negative.

Equation (4.15) constitutes the backbone of the theory of thermodynamics of irreversible processes. From it we deduce that whenever irreversible processes evolve in a system, there is a positive production of local entropy. This can be evaluated by multiplying all flows by their corresponding forces, thus forming a bilinear expression.

The flows and forces described in Fig. 4.3 are not the only entropy-producing processes. They have been derived for systems with only thermal, diffusional, and chemical forces. However, if other contributions to the internal energy, such as electrochemical potentials, exist, they must be added to the Gibbs equation, and thus introduce extra terms of the form $X_{ele} J_{ele}$ into the expression of σ in Eq. (4.15). Let us note that in the literature one sometimes uses the entropy production multiplied by the temperature $\Phi = T\sigma$. The latter is called a *dissipation function*.

We may hope to use Eq. (4.15) in order to get information about a flowing fluid and chemical processes. However, we are trapped in a vicious circle. If we want to calculate σ we need X_i and J_i. On the other hand, if we could evaluate X_i and J_i as functions of time and space, then we would have a complete knowledge of the system and any thermodynamic formalism is superfluous. Unfortunately, the evaluation of forces and fluxes is not an easy task. To make the matter worse, these two interdependent quantities usually appear as complex and unknown functions of each other. Therefore, the formalism of the thermodynamics of irreversible processes consists of the use of Eq. (4.15), taken together with additional assumptions, in order to make predictions about nonequilibrium macroscopic systems. Two different approaches are possible:

1. One finds, by appropriate approximations, simple proportionality relationships between fluxes and forces. This line of study forms the theory of *linear thermodynamics of irreversible processes*.
2. One tries to use Eq. (4.15) without linearization. This constitutes the domain of *nonlinear thermodynamics of irreversible processes*.

4.5. LINEAR THERMODYNAMICS OF IRREVERSIBLE PROCESSES

As we have seen, the central problem of the thermodynamics of irreversible processes is to compute the entropy production of a nonequilibrium system. If in the summation of Eq. (4.15) different forces and fluxes enter neatly by pairs, in the actual physical world they are all interconnected in some unknown and complex fashion. The following examples illustrate this point.

In 1821 Seebeck discovered the *thermoelectric effect*, which can be seen in the following experiment. An antimony wire is connected to a bismuth wire, as shown in Fig. 4.4. If one junction is heated and the other kept cool, an electric

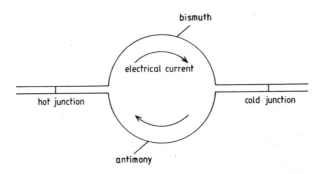

Fig. 4.4. Thermoelectric effect. A temperature gradient generates electrical current.

current flows around the loop in the direction indicated by the arrows. Here a temperature gradient generates an electrical current.

With this example and our daily experience and intuition, we can generalize and say that if a system is subject to different flows and forces, every flow is dependent in some degree on all other forces, and, conversely, each force is related to all flows appearing in the system: $J_i = J_i(X_1, \ldots, X_n)$ and $X_i = X_i(J_1, \ldots, J_n)$. Here n is the total number of forces and fluxes which appear in the expression of entropy production σ. In general, the explicit functional relationships between these quantities are not known. If such relationships could be found in a few special cases, then we would be able to simplify the problem by half and eliminate, for example, all J_i from Eq. (4.15) of entropy production σ.

If the system is in equilibrium, all forces and consequently all corresponding flows vanish. Therefore, it is reasonable to think that if we introduce weak forces or small gradients into the system, the ensuing flows will be proportional to the forces. Also, the properties of the new nonequilibrium system are not drastically different from those of the equilibrium state. The new system remains in the *neighborhood* of the equilibrium state.

In the neighborhood of the equilibrium state, linear relationships between forces and fluxes may be found by an elegant mathematical formulation. As forces are weak in this region, we may expand J_i in function of small X_i, in a Taylor series around the equilibrium state, and consider only first-order terms

$$J_i = J_i^{eq} + \sum_l \left(\frac{\partial J_i}{\partial X_l} \right)_{eq} X_l + \cdots \tag{4.16}$$

By definition all fluxes J_i^{eq} are nonexistent at equilibrium, and therefore,

Eq. (4.16) reduces to a *linear relationship* between flux J_i and all forces X_l

$$J_i = \sum_l \left(\frac{\partial J_i}{\partial X_l}\right)_{eq} X_l \tag{4.17}$$

In this equation the coefficients of proportionality are computed at the equilibrium state. Traditionally they are designated by L and are called *phenomenological coefficients*. With this notation Eq. (4.17) reads

$$J_i = \sum_l L_{il} X_l \tag{4.18}$$

L_{ii} are called *straight coefficients* as they relate a given force to its own flux. When the two indexes of L_{il} are different, it indicates that the i flow is produced by the l force. These are called *coupling coefficients*. Expansions such as Eq. (4.16) are only possible for slow processes, as in this case a small force will produce a small flux.

Even before the development of the thermodynamics of irreversible processes, a few linear phenomenological laws had been known for a long time. For example, in 1822 Fourier discovered that the propagation of heat in solid bodies obeys a simple linear law which relates the thermal flux to the temperature gradient:

$$\mathbf{J}_q = - L_{qq} \frac{\text{grad } T}{T^2} = - \lambda \,\text{grad } T \tag{4.19}$$

$\lambda = L_{qq}/T^2$ is the coefficient of thermal conductivity.

Later on, Fick established that the diffusional flux of matter is proportional to the force which generates the flux. As $\mu_y = \mu_y(\{\rho\})$, one shows immediately that

$$\mathbf{J}_d^\gamma = - L_{dd} \left(\text{grad} \frac{\mu_y}{T} \right)_{Tp} = - D_y \,\text{grad } \rho_\gamma \tag{4.20}$$

Here $D_y = (L_{dd}/T) (\partial \mu_y / \partial \rho_y)_{Tp}$ represents the diffusion coefficient in a dilute mixture.

Approximate linear laws of the form of Eq. (4.18) are quite general and may be written for all common forces and fluxes. The specific nature of matter enters this equation via the phenomenological coefficients L_{il}. For example, in the process of diffusion, there are diffusion coefficients proportional to L_{dd} for each species. However, Fick's law of diffusion may be valid for every diffusing substance. Phenomenological coefficients L_{il} are entirely determined by the internal structure of the medium, and are independent of the forces applied to the system. However, they may be functions of state variables T, p and composition. For example, the diffusion coefficient may increase with concentration, or an increase in pressure may make matter flow more readily. In practical applications it is convenient to assume that the phenomenological coefficients are constant. The important fact remains that they are not explicit functions of forces or fluxes operating in the system.

Let us remember that our final aim is to study the irreversible phenomena in a fluid. In particular, we might want to know how fluxes can be measured when different forces are applied. Equation (4.17) is a great step toward this aim. Unfortunately, we have introduced many unknown phenomenological coefficients L_{li} into the problem. The latter are well-defined equilibrium quantities, but must still be measured by independent experiments. If there are n flows and n forces in the system, we are left with $n \times n$ phenomenological coefficients that must be determined experimentally. For example, for three forces and three fluxes this number is already nine. Fortunately we may reduce this number substantially with the help of two observations:

1.   There are symmetry requirements on coupling of irreversible processes. Not all forces can be coupled to every flux.
2.   The Onsager symmetry laws state that the coefficient L_{li} which relates flux l to force i is equal to the coefficient L_{il} which relates flux i to force l.

The first statement originates from the observation of physical processes. The second is based on the molecular theory of matter and is an extra-thermodynamic result. In the next section we shall have a closer look at Onsager's symmetry laws, or *reciprocal relations*.

4.6. THE ONSAGER RECIPROCAL RELATIONS

In 1931, relying on the molecular nature of matter, Onsager was able to show one of the most fundamental relations of linear irreversible thermodynamics, namely, that for every substance, the coupling coefficients between forces and flows are related by a reciprocal relation

$$L_{il} = L_{li} \qquad (4.21)$$

This equality reduces the number of coupling coefficients by half, and therefore simplifies experimental problems considerably. A few years after Onsager's original work, Casimir showed the validity of these reciprocal relations for a larger class of irreversible processes. In the fifty years since their formulation, Onsager's reciprocal relations have been tested for a wide range of flows and forces, and their validity seems to be universal. In the 1940s, with the help of these relations Meixner and Prigogine laid down a consistent theory which opened the way for the explicit calculation of a great number of physical quantities. Thus, a new field of macroscopic physics, the *linear thermodynamics of irreversible processes*, was born. Until recently, its use was restricted to the vicinity of the equilibrium state, and thus for a long time the thermodynamics of irreversible processes was synonymous with the linear thermodynamics of irreversible processes.

Once a firm foundation for the theory was laid, it developed rapidly in various directions. In the 1960s De Groot and Mazur published a book in which they applied the theory to concrete physical problems, ranging from chemical reactions, heat conduction, viscous flow, electrical conduction, and polarized matter to discontinuous systems which may be seen as prototypes for biological membranes. A few years later, Katchalsky and his co-workers extended the methods of irreversible thermodynamics into the field of biophysics. More specifically, they developed various models for the study of the permeability problems of biological membranes.

The rest of this chapter is devoted to a few examples illustrating the power of irreversible thermodynamics in the study of physical systems. But first let us show a few consequences of the reciprocal relation, Eq. (4.21).

Let us consider a nonequilibrium thermodynamic system subject to heat and matter flow. The corresponding forces, namely, the temperature and concentration gradients, are sufficiently weak such that the local entropy production may be expressed in a linear approximation according to Eq. (4.16),

$$\sigma = L_{qq} X_q^2 + L_{dd} X_d^2 + (L_{qd} + L_{dq}) X_d X_q > 0 \qquad (4.22)$$

Let us imagine a case whereby in Eq. (4.22) the temperature gradient vanishes, and only X_d is non-zero. We are left with $\sigma = L_{dd} X_d^2 \geqslant 0$. This inequality implies that the straight coefficient L_{dd}, which is the diffusion coefficient, must be positive—a known result that we have found by thermodynamic methods. We can also maintain the temperature gradient but stir the solution in order to erase all concentration inhomogeneities. In this case $\sigma = L_{qq} X_q^2 > 0$. We find again a well-known result, namely, that the coefficient of thermal conductivity L_{qq}/T^2 is positive.

With some simple mathematics and Onsager's reciprocal relation (4.21), one can easily show that the products of straight coefficients are larger or, at most, are equal to the square of coupling coefficients:

$$L_{qq} L_{dd} \geqslant (L_{qd})^2$$

For linear systems and for any arbitrary number of forces and fluxes, one can show quite generally that

$$L_{ii} > 0, \quad L_{ii}L_{jj} \geqslant (L_{ij})^2 \tag{4.23}$$

These relations are of crucial importance in the linear thermodynamics of irreversible processes.

4.7. THERMAL DIFFUSION

We now show how practical information about the state of a system may be extracted from a thermodynamic approach to a nonequilibrium system. All nonequilibrium systems are subject to various flows. For example, matter flows from areas of high concentration toward low-concentration zones; heat flows between two regions of different temperature. Therefore, one of the major problems in this kind of system is to determine the rate at which matter or heat travels. It is also important to know if a given flux generates other nonconjugated forces, determine their nature, and finally measure the coupling coefficients. We shall illustrate this procedure with the following simple example.

A tube is filled with a binary solution of uniform composition, for example, a nonreacting red dye, called *solute*, which is dissolved in a large quantity of water, called *solvent*. The two ends of the tube are sealed. One end of the tube is held in a cold thermostat at temperature T_C and the other end is immersed in a hot thermostat at T_H. After a while the mixture reaches and remains at a steady-state regime. The temperature of the solution shows a time-independent profile as depicted in Fig. 4.5. It is obvious that all the properties of the fluid in the tube are subject to continuous gradients, and consequently they are called *continuous systems*. Let us evaluate the diffusion coefficient of solute relative to the solvent, and also the coefficient of thermal conductivity of the solution for the continuous system. We might also wonder if there are interference effects between the diffusion of matter and the heat flow.

As we have seen, the first step in any thermodynamic study of an irreversible process is the evaluation of entropy production σ. In the present case the entropy production is the sum of three contributing factors: the imposed temperature gradient gives rise to a heat flux, and the gradients in the chemical potentials of the solute and the solvent both generate diffusional fluxes for these species. From Eq. (4.11) and Fig. 4.3 we deduce immediately

$$\sigma = \mathbf{J}_q \cdot \text{grad} \frac{1}{T} + \mathbf{J}_d^s \cdot \text{grad}\left(-\frac{\mu_s}{T}\right) + \mathbf{J}_d^r \cdot \text{grad}\left(-\frac{\mu_r}{T}\right) \tag{4.24}$$

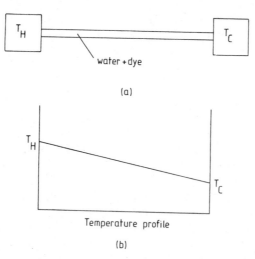

Fig. 4.5. A thermodiffusion cell. (a) A temperature gradient generates a concentration gradient. (b) Temperature profile at steady state.

Here μ_s and μ_r are, respectively, the chemical potentials of the solvent and the solute ($\mu_i = \mu_i(\{C_i\})$).

Before we go any further, let us make some remarks about computational "tricks" that can be used in any theoretical approach. Sometimes a given mathematical expression can be cast into a much simpler form, or a physically more transparent expression, by a change of variables. Let us define another set of forces and fluxes for the thermodiffusional cell of Fig. 4.5. If h_r and h_s are, respectively, the partial molar enthalpies of the solute and the solvent, we consider $\mathbf{J}'_q = \mathbf{J}_q - h_s \mathbf{J}^s_d - h_r \mathbf{J}^r_d$ together with its corresponding force grad $(1/T)$. This new *reduced heat flux* is the difference between the total heat flux $\mathbf{J}_q$ and the heat flows resulting from the flow of matter. If C_r and C_s are respectively the dye and solvent concentrations, $\mathbf{J}'^r_d = \mathbf{J}^r_d - \mathbf{J}^s_d(C_r/C_s)$, defines the flow of solute relative to the solvent. We choose $(\mathrm{grad}/T)\,(-\mu^c_r)$ as the corresponding force of this new flow. Here μ^c_r is the concentration-dependent part of the chemical potential. These new variables are chosen in such a way that the local entropy production σ remains unchanged or *invariant*,

$$\sigma = \Sigma X_i J_i = \Sigma X'_i J'_i \geqslant 0 \tag{4.25}$$

With this transformation we illustrate a most general and important point. In a given system forces and fluxes may be defined in many different ways; the only restriction is that they respect the invariance of σ as defined by Eq. (4.25).

With these new variables and some standard and cumbersome thermodynamic transformations, Eq. (4.24) becomes

$$\sigma = \mathbf{J}_q' \cdot \frac{\text{grad}\,(-T)}{T^2} + \mathbf{J}_d'' \cdot \frac{\text{grad}\,(-\mu_r^c)}{T} \tag{4.26}$$

Comparison of Eqs. (4.24) and (4.26) shows that, by an appropriate mathematical transformation, we have reduced the number of fluxes and forces from three to two. This reduction simplifies the problem considerably. We have seen that three fluxes and forces necessitate nine phenomenological coefficients, whereas Eq. (4.26) may be treated with only four coefficients. It is important to notice that all reference to solvent gradient has disappeared from Eq. (4.26). This is because solvent and solute fluxes are not independent.

We have already said that, in general, all forces and fluxes in a given system may be coupled together. Therefore we write $J_2 = J_2(X_1, X_2)$, and for weak forces we only have to consider a linear relationship between forces and fluxes as given by Eq. (4.17):

$$\mathbf{J}_d'' = -L_{dd}\frac{\text{grad}\,\mu_r^c}{T} - L_{dq}\frac{\text{grad}\,T}{T^2}$$

$$\mathbf{J}_q' = -L_{qd}\frac{\text{grad}\,\mu_r^c}{T} - L_{qq}\frac{\text{grad}\,T}{T^2} \tag{4.27}$$

Let us look in more detail into the physical significance of the phenomenological coefficients L_{ij}. From the definition of these coefficients in Eq. (4.16), we see that L_{dd} expresses the increase in the flux J_d'' consequent to a unit increase of the conjugated force (difference in the chemical potential of the solute) which produces this flow. L_{dq} may be interpreted in the same manner. However, this time it is the temperature gradient, a nonconjugated force, which has increased by a unit and has produced a matter flux. L_{qq} and L_{qd} can be interpreted in the same way for the process of heat flux.

We have already seen that a difference in chemical potential, which results from a difference in concentrations, generates a diffusional flux $\mathbf{J}_d$. Equation (4.27) gives us very important information, namely, that $\mathbf{J}_d$ can also be generated under the action of a temperature gradient. This means that differences in temperature in a system generate concentration variations of the constituents of the fluid. Therefore, we must expect a nonhomogeneous distribution of dye in the tube. As a matter of fact, experience shows that one end of the tube is more deeply colored than the other. In general, in a thermodiffusion cell, some substances are more concentrated in the hot end, whereas other species prefer the cold end. These processes are called *thermal diffusion*.

From the second part of Eq. (4.27), we conclude that not only does heat flow because of a temperature gradient but also the established concentration gradient, due to the thermodiffusional effect, contributes to the heat flux. Moreover, from Onsager's reciprocal relations, Eq. (4.21), we find $L_{qd} = L_{dq}$, which states that the cross effects appear with the same proportionality coefficients. This example illustrates the predictive power of the linear thermodynamics of irreversible processes. However, we can go further and obtain a few quantitative results.

In thermodynamic theory one can transform equations from one set of variables into another set. One reason for this is that thermodynamic variables are a mixture of abstract and quantifiable quantities. Our aim is ultimately to express the equations in terms of measurable quantities. Also, a repeated mathematical transformation may change a given equation into a much simpler form. Now we would like to express Eq. (4.27) as a function of a directly measurable quantity, namely, the concentration of the red dye. One can see that grad $\mu_r^c = (\partial \mu_c / \partial C_r)$ grad C_r. On the other hand, as we have already shown, the coefficients L_{qq}, L_{dd} and L_{qd} are functions of the state of the system. For example, they might vary with the temperature T or the concentration C_r of the solute. In the present problem it has been found that L_{dq} varies linearly with the solute concentration C_r. When these changes are introduced into Eq. (4.27) we get $((\partial \mu_c / \partial C_r) = \mu_{rr})$

$$\mathbf{J}_d'' = - D \operatorname{grad} C - CD^T \operatorname{grad} T$$
$$\mathbf{J}_q' = - \mu_{rr} T C D^D \operatorname{grad} C - \lambda \operatorname{grad} T \qquad (4.28)$$

(We have omitted the index r of C_r as there is only one dye species in the tube.) The new phenomenological coefficients in Eq. (4.28) are defined in the following manner: D is the familiar diffusion coefficient, D^T is the thermal diffusion coefficient, $D^T = L_{dq}/CT^2$, λ is the heat conductivity coefficient, and $D^D = L_{qd}/T^2C$ is the so-called *Dufour coefficient*. From Onsager relations, Eq. (4.21), we deduce that $D^D = D^T$.

If we wait long enough, the system described in Fig. 4.5 reaches a steady state. Heat continues to flow at a steady rate under the influence of the constant temperature gradient, but the matter flux vanishes. The newly established concentration gradient reaches and remains at a steady value. Consequently, from the first part of Eq. (4.28) we can deduce

$$\frac{\operatorname{grad} C}{\operatorname{grad} T} = - \frac{CD^T}{D} \qquad (4.29)$$

The ratio D^T/D is called the *Soret coefficient*. Equation (4.29) relates four independent quantities, the knowledge of any three of which yields the unknown value of the fourth. The diffusion coefficient D can be measured independently. With some hard work grad C and grad T can be calculated by the methods of Section 4.4. Therefore, from Eq. (4.29) we can readily evaluate the value of the thermal diffusion coefficient D^T. In liquids, the latter is of the order of 10^{-8} to 10^{-10} cm^2 sec^{-1} degree^{-1}, whereas the ordinary diffusion coefficient D is of the order 10^{-5} cm^2/sec. Thus the concentration gradient is usually very small unless the temperature gradient is very high.

The determination of the Dufour coefficient D^D follows the same line of thought. This time $\mathbf{J}'_q$ is set at zero in the second part of Eq. (4.28). We find

$$D^D = \frac{-\lambda \operatorname{grad} T}{\mu_{rr} CT \operatorname{grad} C} \tag{4.30}$$

In practice one starts with two different fluids, initially at the same temperature, and lets them diffuse into each other. After some time a temperature gradient appears in the system. For gaseous mixtures one finds temperature differences of the order of one degree centigrade. And, most importantly, it can be verified experimentally that D^T and D^D have the same value, thus proving the validity of Onsager's reciprocity relations. This simple example shows how the methods of linear thermodynamics of irreversible processes may be used to detect the presence of various phenomena in fluid flow and at the same time measure the magnitude of various fluxes. Thermodiffusion phenomena in a binary gas phase is of considerable importance for the separation of isotopes.

4.8. THE CURIE–PRIGOGINE SYMMETRY LAW

The example in the preceding section was extremely simple, since only three phenomenological coefficients had to be considered. However, in most cases the number of coefficients is very large, especially when a few chemical reactions are present. For many systems this number may be reduced considerably by virtue of the Curie-Prigogine symmetry law. Let us go back to the system of Fig. 4.5; but, this time, we mix two substances that can react. In a first approximation, we may neglect all diffusional processes. Therefore, the entropy production takes the simple form (see Fig. 4.3)

$$\sigma = \mathbf{J}_q \cdot \operatorname{grad}\left(\frac{1}{T}\right) + \frac{Av}{T} \geqslant 0 \tag{4.31}$$

Here the chemical rate v is a generalized flux generated by the generalized force A/T which is proportional to the reaction affinity. Therefore the linearized equations are

$$\mathbf{J}_q = - L_{qq} \frac{\text{grad } T}{T^2} + L_{q1} \frac{A}{T}$$

$$v = - L_{1q} \frac{\text{grad } T}{T^2} + L_{11} \frac{A}{T} \qquad (4.32)$$

and introduce four phenomenological coefficients.

Let us suppress the temperature gradient by inserting the tube of Fig. 4.5 in a thermostat; thus grad $T = 0$. From Eq. (4.32) we must conclude that $\mathbf{J}_q = L_{q1}(A/T)$ and is a directional property. However, the nonequilibrium system is close to the equilibrium state and in the expansion, Eq. (4.32), the latter is the reference state and is *isotropic*: the properties of the system are the same in all directions. An isotropic system has the highest possible symmetry in space. If we remember that the phenomenological coefficients L_{ij} are computed at equilibrium state, we conclude that they also are nondirectional, or *scalar* quantities. On the other hand, the chemical force A/T, at a given point, is a measure of the possibility of a chemical process at that point, and is therefore a nondirectional scalar quantity. Therefore, logic and intuition tell us that the product of two scalar quantities $L_{q1}(A/T)$ cannot give rise to a directional heat flux $\mathbf{J}_q$.

Curie (in crystal physics) and Prigogine (in nonequilibrium systems), using general symmetry principles, showed the impossibility of such coupling. The statement is quite general and reads: the causes of different phenomena cannot have more elements of symmetry than the effect they produce. Here the scalar cause A/T is much more symmetrical than the directional effect $\mathbf{J}_q$; therefore it violates the Curie-Prigogine principle. We conclude that $L_{q1} = 0$; but, according to Onsager reciprocal relations, $L_{q1} = L_{1q} = 0$. Therefore the coupling between the thermal effects and chemical processes disappears and σ splits into two independent parts. One part reflects the entropy production due to thermal processes, and the other results from chemical reactions alone. As each term individually may reach any arbitrary magnitude, the only way that their sum can remain positive is that each part separately must be positive. Therefore we have

$$\sigma_q = L_{qq} \left(\frac{\text{grad } T}{T^2} \right)^2 \geqslant 0$$

$$\sigma_r = L_{11} \left(\frac{A}{T} \right)^2 \geqslant 0 \qquad (4.33)$$

With this simple example we have seen that in the linear domain of nonequilibrium thermodynamics, and *only in this domain*, all the irreversible processes are not necessarily coupled. This fact enormously simplifies the actual analysis of the processes. The rule is that only phenomena with the same directional character can influence each other. Scalar properties only couple to other scalar magnitudes, and likewise directional properties are coupled only with each other.

In the nonlinear range, the property of isotropy is lost independently of the structure of the medium at equilibrium. In this case the Curie-Prigogine law is no longer valid. The coupling of chemical reactions and the directional phenomenon of diffusion may lead to "symmetry breaking" or the spontaneous emergence of spatial patterns in a previously uniform medium.

4.9. DIFFUSION

The phenomenon of diffusion of matter is part of our daily experience. In Section 3.11 we have already seen why the smell of perfume spreads in a room. The diffusion process is one of the major agents of self-organization of matter; therefore, in what follows, we shall need a better knowledge of the phenomenon. A quantitative description of diffusion in space and time may be derived from the following example. We take a long, thin tube filled with water at constant temperature T. From a hole in the middle of the tube we inject a drop of ink of mass m into the tube, as shown in Fig. 4.6a. As diffusion is assumed to be the only irreversible process inside the tube, from Fig. 4.3 we conclude that the local entropy production is given by $\sigma = \sum_\gamma \mathbf{J}_d^\gamma \cdot \mathrm{grad}\,(-\mu_\gamma/T)$. With the help of the phenomenological relations given in Eq. (4.18) and the property $(\partial/\partial x)\mu_\gamma = (\partial\mu_\gamma/\partial\rho_\gamma)\,(\partial\rho_\gamma/\partial x)$ we find Fick's law for a one-dimensional system

$$J_d^\gamma = -D_\gamma \frac{\partial \rho_\gamma}{\partial x} \tag{4.34}$$

$D_\gamma = (L_{qq}/T)\,(\partial\mu_\gamma/\partial\rho_\gamma)_{Tp}$ is the *diffusion coefficient* of the ink. In Section 4.4 we showed that the change in time of the density ρ_γ was related to the diffusional flux $\mathbf{J}_d^\gamma$ by Eq. (4.6). This expression combined with Eq. (4.34) and the assumption that $D = $ constant gives

$$\frac{\partial \rho_\gamma}{\partial t} = -\,\mathrm{div}\,\mathbf{J}_d^\gamma = D_\gamma \frac{\partial^2 \rho_\gamma}{\partial x^2} \tag{4.35}$$

and constitutes *Fick's second law* of diffusion. This simple partial differential

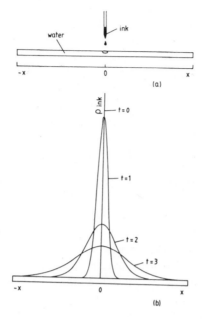

Fig. 4.6. (a) An inhomogeneous system: a drop of ink and water. (b) Diffusion process gradually homogenizes the system.

equation may be integrated easily and yields

$$\rho_\gamma(x,t) = \frac{m}{\sqrt{4\pi D_\gamma t}} e^{-(x^2/4D_\gamma t)} \tag{4.36}$$

Fig. 4.6b shows the ink distribution in water for different times t_0, t_1, t_2, and t_3 as a function of the distance measured from the tip of the needle. From the graph, we see that the ink concentration decreases around the origin and is distributed along the tube such that for long times the entire tube takes a uniform pale blue coloring. The system has reached thermodynamic equilibrium and the irreversible processes have vanished.

4.10. DISCONTINUOUS SYSTEMS: THERMAL OSMOSIS

In the preceding section, we used the methods of the thermodynamics of irreversible processes to study continuous systems where change in properties was gradual throughout the system. However, there are many interesting problems in which the properties change suddenly, in a big jump, at several points inside the system. A tank containing a fluid and partitioned by a semipermeable membrane is an example of such a system (see Fig. 4.7). Much more interesting examples may be found in living organisms. All cells are surrounded by membranes, and inside the cells organelles such as mito-chondria in turn are surrounded by other membranes. As we shall see in Chapter 12, biological membranes are complex structural and functional tissues. However, in a thermodynamic approach, a membrane is essentially a barrier which delimits "outside" from "inside" and creates a discontinuity in space. Waste products and nutrients constantly cross these membranes in both directions and thus the investigation of the permeability of biological membranes may be considered to be of paramount importance. It is therefore desirable to extend the simple methods of the previous sections to the study of these *discontinuous systems*.

We shall consider the simplest possible model which accounts for a transport process across a membrane. Fig. 4.7 represents such a three-phase system. A membrane of thickness Δd separates two compartments, I and II, filled with a solution or a gas mixture. These substances may cross the membrane from one compartment to the other. The system may be broken down into three parts, the inside compartment I, the outside compartment II and a third part, the membrane itself. The methods of Section 4.7 may be extended to each zone separately, considered as continuous systems. The total entropy production of the system is found by adding contributions from the three separate parts. Once the entropy production is found, one can deduce from it all the relevant fluxes and forces acting in the system. From this point on we follow the same procedure as the one outlined for the study of thermodiffusion phenomena. Details of calculations are beyond the scope of the present book.

The problem and its solution can be simplified by a few reasonable

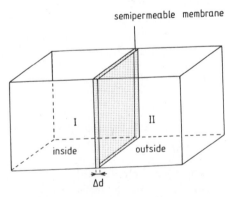

Fig. 4.7.   A three-phase discontinuous system.

assumptions and the appropriate choice of experimental conditions. For example, we may modify the experimental setup of Fig. 4.7 to obtain the one depicted in Fig. 4.8. In the beginning, the two compartments I and II are kept at constant temperature by immersion in two thermostats maintained respectively at temperatures T_1 and T_2. There is no hydrostatic pressure within either compartment. Constant stirring suppresses all inhomogeneities of composition in the two compartments. The membrane is also assumed to be entirely homogeneous. The two compartments and the membrane form a closed system. An inspection of the experimental setup of Fig. 4.8 shows that the fluxes are the same in every point of the membrane surface and are perpendicular to it. This system is the discontinuous version of the thermodiffusional experiment described in Section 4.7. If we compare the two experiments, we see the following fundamental differences:

1. There are no gradients in compartments I and II of the discontinuous system. Gradients are confined to the membrane compartment alone.
2. There are discontinuous jumps in the magnitude of state variables as they cross the membrane from one compartment to the next.

We can simplify the problem still further if we assume that we have waited long enough for a steady state to be established in the system. In such a condition, remembering that both sides of the membrane are homogeneous, we expect that all fluxes across the membrane are constant. The physical significance of this statement is that the same flow crosses a unit area, whether the latter is part of a single phase or separates two phases from each other.

Let us evaluate as usual the local entropy production of the system. Because

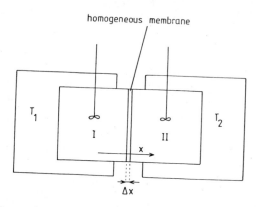

Fig. 4.8. Thermal osmosis. A temperature difference generates a pressure difference. Carbon dioxide is at higher pressure in the hot compartment T_2.

of constant stirring there are no gradients in compartments I and II, and all irreversible processes proceed inside the membrane alone, so that the latter is equivalent to the previous continuous system of Fig. 4.5, but with two differences: the length of the tube has shrunk to Δd and fluxes are assumed constant inside the membrane.

Inspection of Fig. 4.8 and a moment's reflection show that by appropriate alignment of our coordinate system, we can make the x axis perpendicular to the membrane. Therefore, forces in the membrane are only in the direction x, and Δd can be written as $\Delta d = \Delta x$. We write the local entropy production σ within the membrane for a volume element of unit area and of infinitesimal thickness dx. Assuming again the validity of local equilibrium hypothesis for infinitesimal thickness dx, σ is given by Eq. (4.11). A greater simplification follows if no chemical reactions are present inside the membrane and the only processes of importance are heat and matter flows.

If the fluid is a binary mixture, we may reduce the number of diffusional flows to one by defining a flow of solute relative to the solvent as we have already done in Section 4.7. If the membrane separates two gas phases, interference between matter transport and heat flow can be seen even with a one-component gas. Thus, for the sake of simplicity, we consider a one-component gas and choose carbon dioxide (CO_2) as the flowing fluid. (Carbon dioxide is the major component of the "exhaust fumes" of most living organisms.) Moreover, we start the experiment with the same initial pressure or the same chemical potential on both sides of the membrane. With all these simplifications, the local entropy production inside the membrane of Fig. 4.8 takes the simple form

$$\sigma = J_q \frac{d}{dx}\left(\frac{1}{T}\right) + J_d \frac{d}{dx}\left(-\frac{\mu}{T}\right) \tag{4.37}$$

If ΔT and $\Delta \mu$ designate the difference in temperature and chemical potential between compartments I and II, one shows that if $J_q^* = J_q - \mu J_d$, the membrane entropy production is given by

$$\sigma = J_q^* \frac{\Delta T}{T^2} + J_d \frac{\Delta \mu_\gamma}{T} \tag{4.38}$$

It is interesting to notice that in Eq. (4.38) all reference to membrane properties disappears. Only the difference $\Delta \mu$ and ΔT of the two sides of the membrane remain relevant. Therefore ΔT and $\Delta \mu$ appear as new thermodynamic forces. These forces are different from $\text{grad}(1/T)$ and $\text{grad}(-\mu/T)$, but we have already seen that one can always define new forces and fluxes provided the entropy production remains unchanged. The linearized diffusional and thermal fluxes are:

$$J_d = L_{dd}\frac{\Delta\mu}{T} + L_{dq}\frac{\Delta T}{T^2}$$

$$J_q^* = L_{qd}\frac{\Delta\mu}{T} + L_{qq}\frac{\Delta T}{T^2} \tag{4.39}$$

We see that, like the thermodiffusional system, an imposed temperature difference ΔT may establish a difference $\Delta\mu$ in the chemical potential of an initially homogeneous system.

Experimental observations confirm these theoretical findings. If a temperature gradient is established across a membrane immersed in CO_2 gas, the gas flows from one compartment into the next across the membrane. Eventually the flow stops; $J_d = 0$ and a steady state is established with the generation of a pressure difference Δp (which is a measure of $\Delta\mu$) between the two compartments of the system. In the case of CO_2 the pressure difference is about 0.2 percent for each degree in temperature difference and the higher pressure is established in the hot compartment. For other gases the cold compartment may be at the higher pressure.

For an experimental study of the coupling coefficients $L_{qd} = L_{dq}$, it is more convenient to express $\Delta\mu$ in terms of pressure difference Δp. This can be done by using the following relation from equilibrium thermodynamics: $\Delta\mu = -s\Delta T + v\Delta p$. Here s and v are, respectively, the molar entropy and molar volume of the gas. Following the same line of thought as in Section 4.7, at the steady state we readily find that

$$\left(\frac{\Delta p}{\Delta T}\right)_{J_d=0} = -\frac{L_{dq}}{L_{dd}vT} + \frac{s}{v} \tag{4.40}$$

$-L_{dq}/L_{qq}vT$ is called the thermomolecular pressure difference. Many other properties of fluid flow can be studied using the simple system of Fig. 4.8.

4.11. BIOLOGICAL MEMBRANES

The extension of the methods of the linear thermodynamics of irreversible processes to discontinuous systems opened a wide field of applications in the study of biological systems. We have said that such systems are a mosaic of interconnected compartments separated by membranes of various sorts. Therefore the methods of the preceding section can be applied to them straightforwardly in order to evaluate, for example, such quantities as membrane *permeability*, membrane *filtration capacity*, or membrane *conductance*. These are bulk properties that ignore all reference to the molecular processes inside the membrane.

Biological membranes are very different from nonbiological, semipermeable membranes, such as a rubber barrier with holes separating two liquid phases (see Part III). The process of transport of substances across biological membranes cannot be explained by a simple diffusion phenomenon. The diffusional fluxes of these membranes are usually larger than would be predicted by Fick's law alone. Something in the membrane *facilitates the transport.*

In some other cases, biological membranes seem to ignore the fundamental law of diffusion, namely, that matter always diffuses from regions of high concentration into areas of low concentration. In these membranes, matter is transported against its concentration gradient; the phenomenon is called *active transport.* Figure 4.9 depicts a living cell in its environment performing active transport of sodium Na^+ and potassium K^+ ions. These are two key ions in the cell machinery. Potassium ions are more concentrated inside the cell than in the surroundings, while the concentration of sodium ions is higher outside the cell. Therefore the natural tendency of K^+ should be to flow out of the cell, while Na^+ should rush in. Nevertheless, cell membranes are able to "pump" K^+ from outside to inside and simultaneously to eject Na^+ into the surroundings; the two flows are intimately coupled together. Of course there is nothing equivalent to a mechanical pump in the membrane, the countercurrent transport being made possible only by the coupling of chemical reactions inside the membrane with the transport process across the membrane. The fact is well established that at the molecular level chemistry and transport are intimately coupled.

The methods of the linear thermodynamics of irreversible processes have been very fruitful for the study of active as well as facilitated transport and many other properties of biological membranes. Several examples and their detailed theoretical treatment can be found in a book by Katchalsky

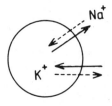

Fig. 4.9. Active transport: living cells transport ions against their concentration gradients (solid arrows). The broken arrows show the direction of the gradients in the absence of active transport.

and Curran. We shall now briefly consider the case of the active transport to illustrate the nature of the coupling between chemistry and transport.

We proceed as in Section 4.10 and write the entropy production of the membrane in the discontinuous system of Fig. 4.8, composed of the three compartments I, II, and III (the membrane). As sodium and potassium are transported across the membrane, matter flux is generated because of the presence of $\Delta\mu_{K^+}$ and $\Delta\mu_{Na^+}$. As the two chemical potentials are related, we can contract these two forces into one which describes the total ion exchange across the membrane, $X_{ex} = \Delta\mu_{Na^+} - \Delta\mu_{K^+}$. Similarily J_{ex} designates the exchange flow of K^+ against Na^+ ions. Using all the resources of the theoretical approach, such as simplifying assumptions and the redefinition of forces and flows, we contract all the relevant chemistry inside the membrane into one chemical flux J_r and its conjugated force X_r. The local entropy production of the membrane is therefore

$$\sigma = X_{ex}J_{ex} + X_rJ_r \tag{4.41}$$

The corresponding linear laws are

$$J_{ex} = L_{11}X_{ex} + L_{12}X_r$$
$$J_r = L_{21}X_{ex} + L_{22}X_r \tag{4.42}$$

Equation (4.42) does not violate the Curie-Prigogine symmetry law, which forbids the direct coupling between transport and chemistry in an isotropic medium. However, a biological membrane is the prototype of a nonuniform or *anisotropic medium*; it is highly selective with respect to the substances it transports. Some substances flow in and others flow out of cells and this directional flow is determined by the anisotropy of the membrane structure itself. Therefore, there is nothing which prevents the coupling of chemistry and transport, in an active transport process, inside an anisotropic membrane. However, the coupling coefficients L_{12}, L_{21} must be directional (vectorial), since they must necessarily reflect the preferential and directional flow of each component.

Active transport, by definition, implies that X_{ex} and J_{ex} are in opposite directions. This is possible only if $L_{12} = L_{21}$ and are different from zero. We have seen that the phenomenological coefficients must satisfy Eq. (4.23), which requires that L_{11}, the straight coefficient, be positive. This in turn implicitly tells us that if $L_{12} = 0$, then J_{ex} and X_{ex} must have the same sign, and active transport becomes impossible. It is the coupling term $L_{12}X_r$ which, by taking negative values, may reverse the direction of J_{ex} from that of X_{ex}. From Eq. (4.41) we see that the active transport introduces a negative contribution $X_{ex}L_{12}X_r$ into the entropy production, thus diminishing the value of the latter. We note that without the coupling of flows and forces of different kinds, entropy-reducing processes would have been impossible in the system.

Active transport is vital for cell life; a cell in resting state uses a third of its "biofuel" to maintain active transport. For example, the maintenance by active transport of a difference in ionic concentrations across the membranes of brain cells confers to the latter electrical properties, vital for the signaling processes in living organisms.

4.12. THE VARIOUS DOMAINS OF THERMODYNAMICS

Let us consider again the synthesis of ammonia from the equation $N_2 + 3H_2 \underset{k_2}{\overset{k_1}{\rightleftharpoons}} 2NH_3$. The affinity A, the generalized force of this reaction, is given by Eq. (3.38). The generalized flux or the reaction rate, following the methods of Chapter 2, is given by the difference between direct and inverse reaction rates

$$v = v_d - v_i = k_1 [N_2] [H_2]^3 - k_2 [NH_3]^2$$

$$v = v_d \left(1 - \frac{k_2 [NH_3]^2}{k_1 [N_2] [H_2]^3} \right) \tag{4.43}$$

Using Eq. (3.40), (3.39), and (2.13) we find

$$v = v_d \left(1 - \exp\left(-\frac{A}{RT} \right) \right) \tag{4.44}$$

Such a nonlinear relationship between fluxes and forces is found for a large class of chemical processes. One verifies immediately that v vanishes as the system reaches chemical equilibrium ($A = 0$ at equilibrium). For states close to equilibrium, the affinity of the reaction is small, and therefore Eq. (4.44) may be expanded into a Taylor's series, and if only the first term is considered,

$$v = k_1 [N_2] [H_2]^3 \frac{A}{RT} \tag{4.45}$$

Again, we have found the familiar linear relationship between a flux and its corresponding force.

In all the examples treated in the previous sections, the approximate linear laws which relate forces to fluxes were quite adequate for the description of heat and fluid flow. However, in a chemical reaction the affinity is a measure of the eagerness of reactants to form products, and thus a low value of A indicates a low reactivity. For ordinary chemical reactions which progress with a reasonable speed, the linear law is no longer valid and the full nonlinear expression, Eq. (4.44), must be considered. As we shall see later, precisely such nonlinear laws give rise to self-organization in chemical systems.

At present we do not have a general thermodynamic theory for the computation of fluxes and forces in the nonlinear domain. We therefore pass immediately to a new chapter where we show the predictive value of thermodynamic methods.

Chapter 5

EVOLUTIONARY CRITERIA

5.1. INTRODUCTION

In studying the properties of bulk matter from the point of view of thermodynamics, it is sometimes possible to guess the direction of evolution of the thermodynamic processes. This predictive capability of thermodynamic theory is of great value to engineers and scientists, as it helps them in choosing the right experiments and avoiding dead ends. For example, evolutionary criteria have convinced engineers, once and for all, that the heat content of the oceans cannot be used as the only source of power for running a boat.

It is common knowledge that heat flows from hot to cold regions, and this fact may be formulated in terms of an evolutionary criterion. In so many daily tasks, we use one or another of the evolutionary criteria of thermodynamics which have become incorporated in our culture as the "things we know."

The evolutionary criteria and the concept of the stability of thermodynamic systems outlined in the next chapter are intimately related. An evolving system reaches a final destiny if the new state is stable, that is, if a slight disturbance will not make it evolve still further. Usually stability and evolution are accounted for in a single chapter, since the thermodynamic relations and properties they use are sometimes quite similar and in some cases identical. However, we have chosen to devote separate chapters to each topic in order to isolate the physical significance of each concept. Moreover, we feel that in this way the reader may see more clearly the new progress achieved in the theory of nonlinear and nonequilibrium processes, which has been mainly in the areas of the stability and evolution of thermodynamic systems.

5.2. EVOLUTIONARY CRITERIA FOR EQUILIBRIUM STATES

Let us consider a class of thermodynamic systems subject to suitable boundary conditions that eventually reach an equilibrium state. A number of reacting chemicals thrown in a vessel and left alone, or a cold body inserted in a hot thermostat, are examples of such systems. The second law of thermodynamics, Eq. (3.13), predicts that the entropy of these systems increases, $d_i S > 0$, and reaches a maximum value when thermal or chemical equilibrium is established, $d_i S = 0$. Consequently, all equilibrium systems are characterized by a universal evolutionary criterion expressed as

$$d_i S \geqslant 0 \tag{5.1}$$

This criterion predicts unambiguously the direction of evolution of any thermodynamic system: an irresistible drive toward the equilibrium state.

The evolutionary criterion, Eq. (5.1), is of paramount importance, in equilibrium as well as in nonequilibrium thermodynamics. In equilibrium thermodynamics this criterion may be used in order to predict which experiments are possible and which are not. The following two examples illustrate this point.

5.2.1. The Impossibility of Making a Magic Boat

There is a lot of thermal energy available in the atmosphere and in lakes and oceans, especially in the equatorial regions of our planet. Why then do we have to import oil from the Middle East to run a boat or an airplane in Florida? We might think this is due to the lack of appropriate technology and hope for a day when cheap energy can be extracted from the oceans. Unfortunately, a simple calculation based on Eq. (5.1) shatters our hopes.

To see this, let us imagine a magic boat able to extract heat from the surrounding water and use it for a trip around the lake, as shown in Fig. 5.1. Lake I and boat II taken together form an isolated system M, in which W work, has been performed. We assume that the lake is almost at constant temperature T and acts as a heat reservoir. We expect the boat to extract heat from the water and transform it into mechanical work of motion performed over a cycle. From the definition of a state function, we know that whenever work is performed over a cycle, the system returns to its original state and all the state functions remain unchanged. This is the case for state functions,

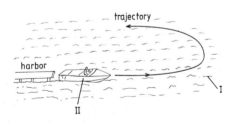

Fig. 5.1. The boat and the lake form the thermally isolated system.

energy, and entropy of the boat. Therefore

$$\Delta U_{II} = 0 \quad \text{and} \quad \Delta S_{II} = 0 \tag{5.2}$$

From the first law, Eq. (3.6), we get a most encouraging result. If W_{II} is the work performed and Q_{II} is the heat received by the boat,

$$\Delta U_{II} = Q_{II} - W_{II} = 0$$
$$W_{II} = Q_{II} \tag{5.3}$$

which implies that the magic boat transforms the totality of Q heat taken from the water into W useful work to move the boat.

However, M is thermally isolated, and therefore, since the lake is the only heat source,

$$Q_{II} = -Q_{I} \tag{5.4}$$

This relation implies that Q_I heat has been lost by the lake: therefore $Q_I/T < 0$, which in turn gives

$$\Delta S_M = \Delta S_{II} + \Delta S_I = \Delta S_I < 0.$$

This fact contradicts the second law, therefore heat cannot flow from the lake to the boat without producing some other effect. Thus the evolutionary criterion, Eq. (5.1), has also been violated. To our great disappointment, a magic boat cannot be constructed, since the heat does not flow naturally from the lake to the boat's engine and produce an equivalent amount of mechanical work. Our conclusions are the consequence of the division of the thermally isolated system M into two subsystems, the boat and the lake, and their interaction.

The example of the magic boat shows the predictive capability of a theoretical approach. With the help of an abstract concept, such as entropy, and a few simple mathematical transformations, we are able to avoid the dead-end alleys in the understanding and exploitation of nature.

5.2.2. A Real Boat

Now let us see if we can account for the functioning of a real boat. We can better understand how one works if we consider another isolated system M, the boat engine, formed from three subsystems acting upon each other, thus not violating the second law. This system is shown in Fig. 5.2. A cold part R_1 is held at a constant temperature T_C and a hot part R_2 is immersed in a hot

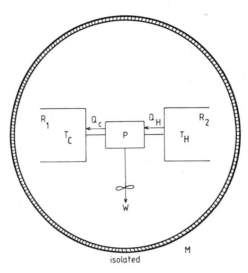

Fig. 5.2. The isolated system is made of the boat engine P inserted between a cold and a hot reservoir.

reservoir T_H. A complex machinery P, the engine, the third part, is held between T_C and T_H. P extracts, in a cyclic operation, Q_H heat from T_H and supplies Q_C heat to T_C. One part of the heat is absorbed and transformed into mechanical work W, which sets the propeller of the boat in motion.

As the system M is isolated, from the preceding example we know that $\Delta S_{R_1} = Q_C/T_C$ and $\Delta S_{R_2} = -Q_H/T_H$. Therefore as $\Delta S_P = 0$

$$\Delta S_M = \Delta S_{R_1} + \Delta S_{R_2} = \frac{Q_C}{T_C} - \frac{Q_H}{T_H} \geqslant 0 \tag{5.5}$$

This inequality may be satisfied if Q_H heat is taken from the hot reservoir T_H and is partly transformed into useful mechanical work W.

If the engine has performed one complete cycle, $\Delta U_P = 0$, and if the process is reversible $\Delta S_M = 0$. As Q_C represents the heat leaving the engine, it has a negative value, which implies in turn that $W = Q_H - Q_C$. Following Carnot, we can define the efficiency of an engine P as $\eta = W/Q_H$. Moreover, one can easily show that $Q_C/Q_H = T_C/T_H$. Thus, the efficiency of a Carnot engine is

$$\eta = 1 - \frac{T_C}{T_H} \tag{5.6}$$

Therefore the further away T_C is from T_H, the more efficient the engine.

Moreover, in Eq. (5.6) all reference to any specific engine has disappeared. Let us note that if $T_C = T_H$, then the system M is equivalent to the system of Fig. 5.1 in which the engine is in contact with a single source of heat. In this case, from Eq. (5.6) we deduce that $\eta = 0$ and the engine stops running.

5.3. THE EVOLUTIONARY CRITERION FOR NONEQUILIBRIUM SYSTEMS—LINEAR DOMAIN

Our preceding examples showed how an evolutionary criterion may be useful in our quest for the secrets of nature. Therefore, it is also desirable to find an evolutionary criterion for nonequilibrium states. However, once fluxes and forces appear, it is no longer possible to make general statements about nonequilibrium systems. The behavior of these will depend critically on the nature and magnitude of existing flows and forces. Nevertheless, the steady states of nonequilibrium systems have special properties, and may be characterized by evolutionary criteria. Let us remember that a system subject to constant and time-independent forces, after an initial turmoil, reaches and remains in a steady state such that no parameter of the system, including entropy, is time-dependent. The thermodiffusion apparatus of Fig. 4.5 and a chemical reactor subject to constant input and output of reactants and products are examples of such systems.

We have seen that in nonequilibrium systems, irreversible processes create entropy

$$\frac{\partial S}{\partial t} = \frac{\partial_e S}{\partial t} + \frac{\partial_i S}{\partial t} \tag{5.7}$$

At steady state $\partial S/\partial t = 0$, which in turn implies that $-d_e S = d_i S$. This equality shows that all increase in the entropy of a nonequilibrium system is compensated by a flow of entropy from the surroundings. This in turn implies that steady nonequilibrium states can only be maintained if energy or matter flows into the system. However, these statements are too general to be of any practical use. We must elaborate a more concrete evolutionary criterion, expressed in terms of relevant forces and fluxes of the nonequilibrium system. In order to achieve this aim, we must distinguish between steady states close to and far from thermodynamic equilibrium, as their evolutionary criteria are not similar. The remaining part of this section will be devoted to the linear systems.

5.3.1. Theorem of Minimum Entropy Production

The evolutionary criterion for equilibrium states was derived with the help of the entropy function S. On the other hand, nonequilibrium systems are characterized by an entropy production P. The latter gives an image of all irreversible happenings in the system expressed in terms of the flows and forces that generate entropy. Therefore, we expect that for nonequilibrium systems, if evolutionary criteria exist, they must be related to the entropy production function P.

Indeed, in 1946 Prigogine showed that the entropy production of a nonequilibrium system not far from equilibrium takes its minimum value at steady states. We prove the *theorem of minimum entropy production* for steady states in the following manner. Let us consider a system subject to n time-independent forces $X_1, \dots, X_n$ such that $X_1, \dots, X_k$ of these forces are kept at fixed values by experimental conditions and give rise to $J_1, \dots, J_k$ non vanishing constant flows. The remaining $X_k, \dots, X_n$ forces are free to adjust to their natural values compatible with the possibilities of the system. In general, these forces have nonzero values, but their corresponding fluxes vanish and we have $J_k = J_{k+1} = \cdots = J_n = 0$. We have already seen an example of such a system in the thermal diffusion experiment of Fig. 4.5. In that example, we had one fixed force, grad $T = X_1$, which gave rise to a nonvanishing steady flow of heat $\mathbf{J}_q = J_1$. The latter established a concentration gradient, grad $C \simeq X_2$. However, as the system was at steady state, the corresponding flux $\mathbf{J}_d = J_2$ vanished. The entropy production of the system is given by Eq. (4.11)

$$\sigma = J_1 X_1 + J_2 X_2$$

As the system is close to the equilibrium state we can use the expansion of Eq. (4.17) and write $J_1 = L_{11} X_1 + L_{12} X_2$ and $J_2 = L_{22} X_2 + L_{21} X_1$. Moreover, with the help of the Onsager reciprocal relations, Eq. (4.21), we find

$$\sigma = L_{11} X_1^2 + L_{22} X_2^2 + 2L_{12} X_1 X_2 \geqslant 0 \tag{5.8}$$

The temperature gradient X_1 is fixed experimentally; therefore only X_2 can adjust its value freely, and thus is a variable of the system. This implies that σ is only a function of X_2, and its derivative with respect to this variable is given by

$$\frac{\partial \sigma}{\partial X_2} = 2(L_{21} X_1 + L_{22} X_2) = 2J_2 \tag{5.9}$$

However, at steady state $J_2 = 0$, and therefore $\partial\sigma/\partial X_2 = 0$ and the entropy production σ has an extremum which is a minimum as Eq. (5.8) is a positive definite function. Therefore, in the domain of the validity of the linear thermodynamics of irreversible processes, the steady states are characterized by a minimum entropy production. Consequently, the system reaches and remains in a state of least dissipation of free energy.

The theorem of minimum entropy production embodies the evolutionary criterion we were looking for. It indicates the direction of evolution of a thermodynamic system kept out of equilibrium by weak and constant forces. As the system is not allowed to reach the ideal state of minimum dissipation, that is of maximum and constant entropy, it chooses the next best alternative and remains in a state of minimum dissipation or minimum entropy production. Let us note that, for linear systems, the entropy production serves the same purpose as thermodynamic potentials in the equilibrium domain.

5.4. THE EVOLUTIONARY CRITERION FOR STEADY STATES FAR FROM EQUILIBRIUM

When a system is far from thermodynamic equilibrium, the dependence of fluxes upon forces is very complex and difficult to evaluate. In general, the magnitude of forces and therefore the corresponding fluxes are large and an expansion of the type given in Eq. (4.18) is no longer possible. Consequently, linear relationships are not available, and the procedure of Section 5.3 cannot be used. With all these difficulties, it still is possible to derive an evolutionary criterion for such far from equilibrium states. We start with the total *entropy production*

$$P = \int dV\sigma = \int dV \sum_k J_k X_k \tag{5.10}$$

If we go back to the theorem of minimum entropy production, we see that the proof relied heavily on the fact that flows could be expressed in terms of forces. Consequently, all fluxes were eliminated from the expression of entropy production. For systems far from equilibrium we are not allowed to make such substitutions. But if in some way we were to separate in the expression of time change of entropy production, the contributions from the rate of change of the thermodynamic forces from the remaining part of dP/dt, maybe some theorem could be demonstrated. This is exactly what Glansdorff and Prigogine did in 1954, deriving an evolutionary criterion for systems far from thermodynamic equilibrium.

Start with the evaluation of dP/dt. As both forces and fluxes are functions of time, and the operation of integration over the volume and differentiation with respect to time are independent,

$$\frac{dP}{dt} = \int dV \sum_k J_k \frac{dX_k}{dt} + \int dV \sum_k X_k \frac{dJ_k}{dt} \tag{5.11}$$

which in short notation may be written as

$$\frac{dP}{dt} = \frac{d_X P}{dt} + \frac{d_J P}{dt} \tag{5.12}$$

Thus the change in time of entropy production splits into two contributions. $d_X P/dt$ shows the change of entropy production due to the rate of change of the thermodynamic forces while the fluxes remain constant, while $d_J P/dt$ results from the rate of change of fluxes while the forces are held constant.

A close scrutiny shows that $d_J P/dt$ does not have any remarkable or general property that can be used as an evolutionary criterion. Fortunately, $d_X P/dt$ has an analogous behavior to dP/dt in the linear domain, namely, that for every thermodynamic system far from equilibrium we have

$$\frac{d_X P}{dt} < 0 \qquad \text{away from steady state}$$

$$\frac{d_X P}{dt} = 0 \qquad \text{at steady state} \tag{5.13}$$

We shall demonstrate these properties with the simple example of heat conduction, in a nonexpanding solid. If a thin metalic rod is inserted between the two reservoirs of Fig. 4.5 at steady state, the local entropy production is given by

$$\sigma = \mathbf{J}_q \cdot \text{grad} \frac{1}{T}$$

Therefore, from Eq. (5.12) we have

$$\frac{d_X P}{dt} = \int dV \mathbf{J}_q \cdot \frac{\partial}{\partial t} \text{grad} \frac{1}{T} \tag{5.14}$$

By the standard technique of integration by parts, one shows straightforwardly that for fixed boundary conditions, the surface term vanishes and we get

$$\frac{d_X P}{dt} = \int dV \frac{1}{T^2} \left(\frac{\partial T}{\partial t} \right) \text{div} \, \mathbf{J}_q \tag{5.15}$$

If the heat flux is the only irreversible phenomenon that proceeds in the rod, we have already shown that the change in the internal energy is given by Eq. (4.9).

Moreover, if ρ is the quantity of matter per unit volume, u the energy per unit mass, c_v the specific heat at constant volume per unit mass, and dT the infinitesimal change in the temperature during the time interval dt, then the rate of change of the internal energy is given by

$$\frac{\partial \rho u}{\partial t} = -\operatorname{div} \mathbf{J}_q = \rho c_v \frac{\partial T}{\partial t} \tag{5.16}$$

In this derivation we made use of the fact that ρ is constant for a nonexpanding rod, and c_v is an intrinsic property of matter and is always time-independent. If we introduce Eq. (5.16) into Eq. (5.15) we find

$$\frac{d_x P}{dt} = -\int dV \rho \frac{c_v}{T^2}\left(\frac{\partial T}{\partial t}\right)^2 \tag{5.17}$$

All squared quantities in this expression are positive, the density ρ is also positive, and the same is true for c_v of the metalic rod. Therefore the integral, Eq. (5.17), is positive, since it is the sum of a product of positive quantities. We thus see immediately that

$$\frac{d_x P}{dt} < 0 \qquad \text{away from the steady state.}$$

By definition, at steady state $\partial T/\partial t = 0$; therefore, from Eq. (5.17) we deduce immediately

$$\frac{d_x P}{dt} = 0 \qquad \text{at steady state}$$

Equation (5.13) is a general evolutionary criterion for nonequilibrium systems, showing the direction of evolution of the thermodynamic system. It applies to all systems for which the entropy production may be evaluated with the help of the local equilibrium hypotheses and which are subject to fixed boundary conditions.

Subsequently, we shall be interested mainly in systems that are characterized only by reacting and diffusing substances. Therefore in the expression of entropy production we shall only keep fluxes and forces relative to these two processes. Then one can show that

$$\frac{d_x P}{dt} = -\frac{1}{T}\int dV \sum_{ij}\left(\frac{\partial \mu_i}{\partial \rho_j}\right)\frac{\partial \rho_i}{\partial t}\frac{\partial \rho_j}{\partial t} \tag{5.18}$$

Naturally the evolutionary criterion, Eq. (5.13), must also be valid in the

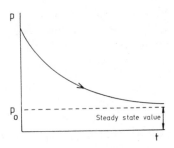

Fig. 5.3. In the linear range, entropy production decreases asymptotically to a minimum value at the steady state.

range of the linear thermodynamics of irreversible processes. This may be verified immediately if fluxes are expanded as linear functions of forces and Onsager's reciprocal relations are considered. One finds that

$$\frac{d_X P}{dt} = \frac{d_J P}{dt} = \frac{1}{2}\frac{dP}{dt} \leqslant 0 \tag{5.19}$$

Figure 5.3 shows the time variation of P as given by Eq. (5.19). We see that under the influence of constant and imposed forces, $P(t)$ decreases in time till a constant minimum value is reached for infinitely long times. This in turn means that entropy production evolves asymptotically and reaches a minimum value at the steady state.

In the linear range, dP is the differential of the function P. This is not the case for $d_x P$. If we could show that $d_x P$ is the differential of a state function Φ, then the latter will have a minimum value at the steady state. In general such a function cannot be constructed. As we shall see in Chapter 6, this lack of a potential function in most nonlinear problems prevents a class of thermodynamic systems from remaining in stable steady states, consistent with the prescribed time-independent boundary conditions of the system.

*   *

*

Presently we turn to the third aspect of thermodynamic theory: we must find a procedure by which the stability of a given state can be assessed.

Chapter 6

STABILITY CRITERIA

6.1. INTRODUCTION

Thermodynamic methods may be used in a third way in order to get information about the behavior of a system around a given state. We may wish to know how a system reacts to spontaneous disturbances, either those within, present in every system, or those imposed from the outside world. Such an analysis may be performed in the framework of the stability theory of thermodynamic systems, and is mainly of a qualitative nature.

The concept of stability is best illustrated with the following example. It is possible, although difficult, to keep a coin standing on its edge on a table. But the slightest knock on the table throws the coin into a new position in which it lies face up or face down. The standing position of the coin was *unstable*, and the slightest disturbance made the system leave this position. On the other hand, the horizontal position of the coin is *stable*, since no matter how much we shake the table the coin will remain face up or face down and will not return spontaneously to the unstable vertical position.

Every thermodynamic system is subject to various spontaneous disturbances or *fluctuations*. For example, the temperature of the system may increase or decrease spontaneously by some arbitrary small amount, noted as δT. As all thermodynamic potentials are temperature-dependent, the system reaches a new state defined, for example, by the entropy $S^{\cdot} = S(V + \delta V, T + \delta T, n)$. With the help of thermodynamic stability theory we investigate the transient or permanent character of this new state. If the system is stable, the fluctuations regress and the entropy returns to its initial value, $S = S(V, T, n)$. On the other hand, if the original state is unstable, a small fluctuation assumes a macroscopic scale and drives the system into a completely new state, distinct from the initial one. Instabilities may appear due to fluctuations of any thermodynamic parameter.

The purpose of thermodynamic stability theory is to derive methods by which the stability concept may be expressed in terms of flows and forces of the system under consideration.

6.2. STABILITY OF EQUILIBRIUM STATES

Let us consider two thermostats, one at $-5°C$, the other at $+60°C$. We introduce two identical glasses of water into each thermostat and wait for thermal equilibrium to be established. Soon we get hot water at $60°C$ in one thermostat, and in normal conditions the glass of water in the second thermostat transforms into ice. However, a physicist can manage, by a rapid cooling of water, to keep the latter at liquid state even at $-5°C$ below the freezing temperature. This abnormal water condition is called the supercooled state. Now we isolate and introduce a small piece of ice (the finite perturbation) into each glass of water. The ice melts in the hot water instantaneously and the system soon returns to its original thermal equilibrium state of $60°C$. On the other hand, the supercooled water transforms immediately into solid ice. Thus we conclude that hot water in thermal equilibrium is in a thermodynamic *stable state*, whereas the supercooled water is in a state of *metastable equilibrium*. Therefore, a finite perturbation brings the system into a more stable state, namely that of ice.

Let us investigate the stability of the glass of hot water in terms of well-defined physical properties of the thermodynamic system. We put the isolated system in contact with a device that is able either to add or extract a small but otherwise undetermined quantity of heat from the glass of water. The thermal device functions randomly and in an unpredictable fashion, as shown in Fig. 6.1. If we only consider thermal effects, then the addition of a small

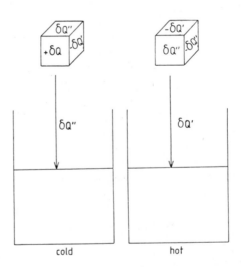

Fig. 6.1.   Random disturbances in a thermodynamic system.

amount of heat δQ changes the internal energy of the system by a quantity $\delta U = \delta Q (\delta V = 0$ for liquids) and the entropy of the system $S_e(U)$ jumps to a new state. As the fluctuating small quantity δU is a random event operating with equally random magnitude and direction, we must distinguish it from the infinitesimal change dU which has a well-defined meaning in differential calculus.

Let us expand the entropy of the system around the equilibrium state

$$S = S_e + \left(\frac{\partial S}{\partial U}\right)_e \delta U + \frac{1}{2}\left(\frac{\partial^2 S}{\partial U^2}\right)_e (\delta U)^2 + \cdots$$

$$= S_e + (\delta S)_e + \tfrac{1}{2}(\delta^2 S)_e + \cdots \tag{6.1}$$

As entropy is maximum at equilibrium state the first-order term $(\delta S)_e$ vanishes, and therefore we need only to evaluate the second-order contribution $(\delta^2 S)_e$. If the latter is negative, then entropy function has a maximum value and the equilibrium state is stable.

Remembering that $C_V = (\partial U/\partial T)_V$ we find immediately that

$$(\delta^2 S)_e = \frac{\partial^2 S}{\partial U^2}(\delta U)^2 = \frac{\partial(1/T)}{\partial U}(\delta U)^2 = -C_V \frac{(\delta T)^2}{T^2} < 0 \tag{6.2}$$

This result is only valid for uniform systems; heterogeneous media need a more detailed study.

We know by experience that if we add heat to the water, the temperature rises, and thus $C_V = dQ/dT$ of normal water is a positive quantity. Therefore, $(\delta^2 S)_e$ is negative for positive and negative fluctuations alike, and we may conclude that the equilibrium state of the hot glass of water is stable. Thus the relation $(\delta^2 S)_e < 0$ may be used as a stability criterion for the investigation of the state of the system (see Fig. 6.4).

A similar treatment may be applied to the supercooled water. The reasoning, however, is more involved and will not be reproduced here.

For an arbitrary uniform system, if several variables are perturbed simultaneously, using Eq. (3.29) one shows a general stability condition

$$(\delta^2 S)_e = -\int dV \frac{\rho}{T}\left[\frac{c_v}{T}(\delta T)^2 + \frac{\rho}{\chi}(\delta v)^2_{N_i} + \sum_{ij}\mu_{ij}\delta N_i \delta N_j\right] \leqslant 0 \tag{6.3}$$

Here c_v is the specific heat at constant volume per unit mass, χ stands for the isothermal compressibility coefficient, μ_{ij} is defined as $(\partial\mu_i/\partial N_j)_{pT}$, and $\delta N_i \delta N_j$ are composition variables. All these quantities can be measured experimentally. The expression under the integral sign is a quadratic form, the only way the stability condition Eq. (6.3) can be satisfied is that the sum of the terms in the bracket becomes positive. Moreover, as these increments are arbitrary in each point, we can imagine instances where only δT, δv, or $\delta N_i \delta N_j$ are present in the system. Consequently every term in the bracket must separately be positive. The only way this may happen is

$$c_v > 0, \quad \chi > 0 \quad \text{and} \quad \sum_{ij} \mu_{ij}\delta N_i \delta N_j > 0 \tag{6.4}$$

These three inequalities constitute the stability condition of equilibrium states, expressed in terms of well-defined measurable quantities, such as c_v, χ, and chemical potentials of the thermodynamic system.

6.3. STABILITY OF STEADY STATES—LINEAR DOMAIN

We have seen that the linear domain of the thermodynamics of irreversible processes is defined as a state not far from equilibrium, but nevertheless subject to flows and forces which are assumed to be small. We have seen also that the steady states of such systems are characterized by a minimum rate of entropy production. Presently we are concerned with the stability properties of these steady states.

We assume that the time-independent steady state fluxes J_i° and forces X_i° are perturbed momentarily by internal or external fluctuations and are given by $J_i = J_i^\circ + \delta J_i$ and $X_i = X_i^\circ + \delta X_i$. Here δX_i and δJ_i are the random fluctuations that appear in the system. Fluctuations drive the system into a time-dependent regime characterized by an entropy $S = S(t)$ and a time-dependent entropy production $P = P(t)$.

With the help of qualitative arguments, it is possible to show that in the course of time all fluctuations disappear and the system returns to the stable steady state characterized by a minimum entropy production.

This can be seen if we go back to the problem of diffusion in its discontinuous version of Section 4.10. In that problem $\Delta T = X_1^\circ$ was the fixed force and gave rise to a constant $\mathbf{J}_q = J_1^\circ$ heat flux. The nonrestricted force $\Delta p = X_2^\circ$ and the corresponding flux $\mathbf{J}_d = J_2^\circ = 0$ were also present at the steady state. Let us assume that some external or internal fluctuations change the force X_2° to a new value $X_2^\circ + \delta X_2$. This force in turn will generate a small flow δJ_2. Thus it is legitimate to write, to a first approximation, a linear relationship between the flux δJ_2 and the fluctuating force that generates it. Thus $\delta J_2 = L_{22}\delta X_2$. The constant coefficient $L_{22} > 0$ must be positive as seen from Eq. (4.23).

The part of entropy production generated only by fluctuating quantities is given by

$$\delta_X P = L_{22}\delta X_2\delta X_2 = L_{22}(\delta X_2)^2 > 0$$

and is a positive quantity. However, as $J_2^\circ = 0, \delta J_2 = J_2 = L_{22}\delta X_2$, and therefore, $\delta_X P = J_2\delta X_2 > 0$. This inequality is satisfied only if the flow J_2 caused by the perturbing force δX_2 has the same sign as the force itself. This fact implies that if, in the thermodiffusion cell, a fluctuation increases the pressure difference Δp above its steady-state value, immediately matter flows from high-pressure regions into the low-pressure areas in order to smooth out the excess pressure difference. The system returns to the stable steady state characterized by a minimum value of entropy production.

For a more formal demonstration of the stability of steady states in the linear domain, we return to the example of heat conduction in a metallic rod inserted between two heat reservoirs. It is easily shown that in the linear range the entropy production of the system is given by

$$P = \int dV L_{qq}\left(\operatorname{grad}\frac{1}{T}\right)^2 > 0 \tag{6.5}$$

The time change dP/dt of the entropy production Eq. (6.5) may be evaluated straightforwardly and exactly in the same manner as $d_X P/dt$ was derived in Eq. (5.17), and one finds

$$\frac{dP}{dt} = -2\int dV \rho\frac{c_v}{T^2}\left(\frac{\partial T}{\partial t}\right)^2 \tag{6.6}$$

The expression under the integral sign describes the local events in each small segment of the rod in the nonequilibrium system. Our main assumption in developing irreversible thermodynamics was that equilibrium relations are valid locally. Therefore, we can use the first relation of Eq. (6.4), which gives the stability of the equilibrium state, and find that c_v is positive. Moreover, as ρ and the remaining terms under the integral sign are also positive, we deduce that

$$\frac{dP}{dt} < 0 \qquad \text{away from steady state}$$

$$\frac{dP}{dt} = 0 \qquad \text{at steady state} \tag{6.7}$$

With this relation we also demonstrate the theorem of minimum entropy production for a second time. The steady states characterized by a minimum entropy production are stable. If fluctuations create entropy, by virtue of Eq. (6.7) its production rate necessarily decreases in time until the steady state is established again.

If irreversible processes are generated solely by chemical reactions and the diffusion of reacting molecules, one shows easily that

$$\frac{dP}{dt} = -\frac{2}{T^2} \int dV \sum_{ij} \left(\frac{\partial \mu_i}{\partial \rho_j}\right) \frac{\partial \rho_i}{\partial t} \frac{\partial \rho_j}{\partial t} \leq 0 \qquad (6.8)$$

In Eq. (6.7) we recognize the evolutionary criterion of steady states in the linear domain. Therefore, in this range the stability and the evolutionary criteria are identical. This will no longer be the case as the system is driven into the nonlinear domain of irreversible processes.

6.4. STABILITY OF STEADY STATES—NONLINEAR DOMAIN

We have seen that the evolutionary criteria and the stability of equilibrium and linear nonequilibrium systems are intimately related. The existence of evolutionary criteria immediately insures the existence of a corresponding stability criterion for the steady states. The minimum value of thermodynamic potentials of the equilibrium state guarantees the stability of the latter. In the same way the minimum value of another potential function, namely the entropy production, insures the stability of steady states in the neighborhood of equilibrium. These properties are quite general and valid for every thermodynamic system. The fate of the system, characterized by a minimum of dissipation, is sealed once and for all by the time-independent boundary conditions.

Many unsuccessful attempts have been made to relate the evolutionary criterion $d_x P/dt \leq 0$ to a new state function Φ such as $d_x P$ would appear as the differential of Φ, in which case $d_x P = d\Phi \leq 0$. If the construction of such *kinetic potential* functions were possible for every thermodynamic system, it would guarantee the stability of all nonequilibrium steady states. But such a kinetic potential cannot be constructed for every nonlinear thermodynamic system. The nonequilibrium states do not share a unique fate in the face of fluctuations. There is no mechanism, such as a potential function, that guarantees the disappearance of the fluctuations. They may be amplified and change the fate of the system drastically.

A stability criterion for far from equilibrium steady states may be established if we investigate the immediate neighborhood of the states under consideration. We assume that the nonequilibrium system has reached a steady state characterized by the entropy function S° and entropy production function P°. Now any fluctuation δJ_i or δX_i of one or several forces and fluxes of the system brings the entropy value to $S = S(t)$ and entropy production also becomes time-dependent. From the material of the preceding two sections we guess that in a nonequilibrium state, the stability criterion must be deduced from some relationship between the entropy and entropy production. As δJ_i and δX_i are assumed to be small and we are exploring only the immediate neighborhood of steady states; S and P may be expanded in a Taylor series as in Eq. (6.1). However we must remember that the reference state is a steady state, and it can be as far from equilibrium as desired, provided that the local equilibrium hypothesis is still valid. Thus

$$S = S^\circ + \delta S + \frac{1}{2}\delta^2 S + \cdots$$

$$P = P^\circ + \delta P + \frac{1}{2}\delta^2 P + \cdots \tag{6.9}$$

A detailed analysis of the explicit expressions of δS and δP does not reveal any remarkable property that can be related to the stability of the steady states characterized by S°. However, the second-order terms $\delta^2 S$ and $\delta^2 P$ are much more promising.

A close scrutiny and some calculations reveals that $(1/2)\delta^2 P$ is related to that part of the entropy production which is produced from all the fluctuating forces and fluxes in the system and is therefore called *excess entropy production*. We find that

$$\tfrac{1}{2}\delta^2 P = \int dV \sum_k \delta X_k \delta J_k = \delta_X P = \text{excess entropy production} \tag{6.10}$$

The last equality in Eq. (6.10) designates the short-cut notation for excess entropy production.

A moment's reflection tells us that we know something about $(1/2)\delta^2 S$. From the assumption of local equilibrium and the content of Eq. (6.3) we know that the condition of stability of the equilibrium state requires that locally $\delta^2 s \leqslant 0$. Thus for the global system we have

$$\frac{1}{2}\delta^2 S = \frac{1}{2}\int dV \rho \delta_s^2 \leqslant 0 \tag{6.11}$$

With the help of Eq. (6.10) one shows easily that

$$\frac{d}{dt}\tfrac{1}{2}(\delta^2 S) = \delta_X P \tag{6.12}$$

This equation is a most welcome relationship, since it connects $\delta^2 S$, which has a well-defined direction of evolution, to $\delta_X P$, which is evaluated from random fluctuations. Thus we expect that Eq. (6.12) may serve as a stability criterion for nonequilibrium states. This is indeed the case, as will be seen from the following discussion.

If we look at the way $\delta^2 S$ has been evaluated we see that it is nonzero only if fluctuations are nonvanishing; thus it measures deviations from the steady state which originate from disturbances. If $\delta^2 S$ decreases in time (becomes less negative), $(d/dt)\delta^2 S > 0$, the system returns to the reference steady state S° and the latter is stable. On the other hand, if $\delta^2 S$ increases in time (becomes more negative) $(d/dt)\delta^2 S < 0$, then the system will never return to its reference state; consequently, the latter is unstable. Therefore, from Eq. (6.12) we conclude that the stability of nonequilibrium steady states may be deduced from the sign of the excess entropy production, $\delta_X P$. If the latter is negative, the system is unstable, whereas positive values of $\delta_X P$ indicate the asymptotic stability of the steady states. Thus the stability criterion of far from equilibrium states is given by

$$\delta_X P = \int dV \sum_k \delta J_k \delta X_k \geq 0$$

These findings are summarized in Eq. (6.13) and Fig. 6.2.

$$
\begin{array}{lll}
t \geq t_0 & \delta_X P < 0 & \text{unstable reference state} \\
t \geq t_0 & \delta_X P > 0 & \text{asymptotically stable reference state} \\
t \geq t_0 & \delta_X P = 0 & \text{marginal stability}
\end{array} \tag{6.13}
$$

In order to test the stability of a given steady state, one introduces fluctuations δJ and δX into the system, and then evaluates the excess entropy production $\delta_X P$. From the sign of this quantity and the content of Eq. (6.13) the stability of the steady states may be assessed.

The stability criterion, Eq. (6.13), has a simple physical interpretation. Let us consider a local, small-volume element in a nonequilibrium steady state which is under the influence of a force X and a flux J. $\delta_X P = \int dV \delta J \delta X$ may become negative if locally $\delta X \cdot \delta J$ is negative. The only way this might happen is when the fluctuating force δX gives rise to a δJ such that in the presence of

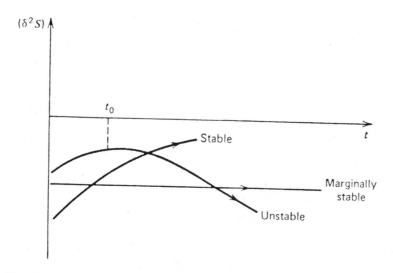

Fig. 6.2. Time change of second-order entropy production $\delta^2 S$ around a nonequilibrium steady state for asymptotically stable, marginally stable, and unstable situations.

vectorial quantities the angle between the two vectors is greater than 90°. This in turn means that an increase in a given force will generate an excess flow in such a way as to increase still further the magnitude of the fluctuating force. Therefore the initial perturbation increases instead of decreasing, thus causing the instability of the steady state. On the other hand, $\delta X \delta J > 0$ implies that the fluctuating force generates locally a flow such that the angle between the two vectors is less than 90°. The fluctuating flux operates in such a way as to make the force disappear; therefore, all fluctuations are damped and the system returns to its initial steady state.

The stability criterion of nonequilibrium states is obtained assuming local equilibrium and the absence of bulk fluid motion. It is only valid in the immediate neighborhood of a reference state. Figure 6.3 shows the domain of validity of the criterion for the case of a simple chemical reaction.

Equation (6.13) is a most general stability criterion valid for all systems obeying the restrictions cited above. Therefore, we must expect that the equilibrium and linear nonequilibrium stability criteria of the preceding sections are special cases of this more general stability criterion. If equilibrium is the reference state, there are no forces and fluxes present, and therefore the perturbations of these quantities are equal to the quantities themselves; thus $J_k = \delta J_k$ and $X_k = \delta X_k$, and the excess entropy production is equal to the entropy production itself, which is always a positive quantity. Moreover we have seen from (6.11) that $\delta^2 S_e \leqslant 0$. Therefore from Eq. (6.13) we deduce

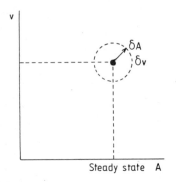

Fig. 6.3. Domain of validity of the thermodynamic stability criterion for a chemical reaction in which v is the rate and A is the affinity of the reaction.

immediately that the reference state is stable, a result we have already shown in Section 6.3. From Eq. (6.12) we see that $(\delta^2 S)_e$ evolves in time, as indicated in Fig. 6.4.

In the linear domain of nonequilibrium thermodynamics, fluxes are linear functions of forces. The same relationship between δJ and δX exists, therefore $\delta J_k = L_{kk}\delta X_k$, and consequently for a one flow system we have

$$\delta_X P = \int dV \delta J_k \delta X_k = \int dV L_{kk}(\delta X_k)^2 \geqslant 0$$

This quantity is always positive according to Eq. (4.23), and therefore the reference steady state is always stable.

If far from equilibrium steady states are not necessarily stable, it means that a given state, if perturbed, will not just adapt to the new constraints by changing slightly all its thermodynamic parameters. On the contrary, fluctuations open new possibilities for energy dissipation, which in turn drives

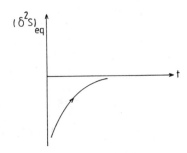

Fig. 6.4. Time change of second-order excess entropy $(\delta^2 S)_{eq}$ around equilibrium state.

the system into other steady states. As the system does exist, at least some of these new states must show stability.

The equilibrium and nonequilibrium systems in linear domain do not exhibit any particular spacio-temporal behavior. If they are driven into the nonequilibrium domain, instabilities may push the system into time-dependent or spatially nonuniform states. A sudden and appreciable change in the properties of the uniform states may also occur.

Today many examples of unstable nonlinear systems are known, and they play a crucial role in the understanding of the dynamic manifestation of matter, the unexpected nature of chemical reactions, the organization of biological systems, and some aspects of living societies from bacteria to man. The rest of this chapter is devoted to several examples of systems characterized by unstable far from equilibrium steady states.

6.5. BENARD OR CONVECTIVE INSTABILITY

The phenomenon we are about to describe was discovered by Benard in 1900, though its diversity and richness have become apparent only recently. Theoretical understanding of Benard-type instabilities is today one of the major fields of research in physics.

Let us consider a thin, horizontal layer of a viscous fluid at rest, as shown in Fig. 6.5. The fluid is maintained between two infinite horizontal plates. The top plate is held at a constant and uniform temperature T_c, while the bottom plate is heated at a constant and uniform temperature $T_H > T_C$. From the material of the preceding section we guess that at the steady state the constant force, $X = ((T_H - T_C)/h) \sim \Delta T$, where h denotes the distance between the two plates, generates a constant heat flux ΔT. Moreover, for small ΔT there is a linear relationship between the force and the flux as seen from the Fourier law, Eq. (4.19). We gradually and slowly increase ΔT such that the fluid may adjust at each moment to the given constraint. We see that, for the critical value ΔT_C, the linear laws, Eq. (4.17), are no longer valid and the system enters a nonlinear domain. Before ΔT_C, the fluid is at rest and looks shapeless, but at ΔT_C a small increase of δX "organizes" the fluid into regular rolls or convection cells of macroscopic size, as shown in Fig. 6.5a. The fluid motion is best visualized if aluminum powder is added to the fluid.

At the molecular level the particles of the fluid move in a cooperative fashion and in unison. Correlations between molecules extend over distances of the order of a centimeter, whereas intermolecular attractive forces act only over distances of the order of 10^{-8} cm. In a circular motion convection cells transfer heat from the bottom to the top of the fluid.

In the uniform fluid, all points of the system are equivalent. However, the

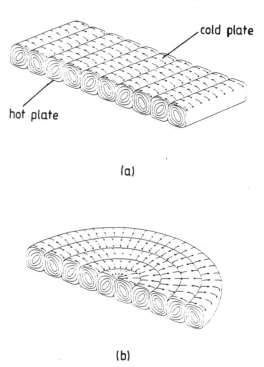

(a)

(b)

Fig. 6.5. A layer of fluid is held between two horizontal cold and hot plates. At ΔT_c stable, roll-shaped convection cells appear. The fundamental unit of the pattern is formed of two oppositely rotating rolls. (From M. G. Velarde and C. Normand, 1980, reprinted by permission *Scientific American.*)

inspection of Fig. 6.5 shows that adjacent rolls rotate in opposite directions. Therefore, locally the symmetry of the fluid has been broken. Equivalent points are only found if one moves a distance of two rolls in the fluid. Therefore we speak of a *symmetry-breaking instability* which result from the energy dissipation in the system.

In a cylindrical container the convection cells are circular rolls, as shown in Fig. 6.5b. One may also observe square, pentagonal, or hexagonal cells. The latter appear if the cold plate is removed from the top of the cylinder. As a general rule, for a given fluid and ΔT_C, a wide variety of patterns may arise whenever the geometrical dimensions of the system are modified.

As ΔT is increased beyond ΔT_C, new self-organized phenomena appear. The sequence of events is not unique and is dependent on the viscosity of the fluid and the dimensions of the container. For liquid helium, heated in a cylindrical container, the following sequence of organized states is seen.

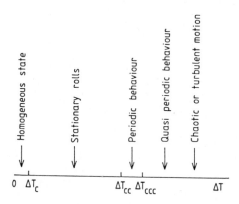

Fig. 6.6. A succession of organized states in liquid helium.

At ΔT_C circular stationary rolls appear and remain until a new critical value $\Delta T_{CC} = 7.4 \Delta T_C$ is reached. At this point, locally the velocity and the temperature of the convection cells change in time periodically, with a constant frequency and in a predictable fashion. This regime prevails until a new threshold $\Delta T_{CCC} = 9.9 \Delta T_C$ is reached. At this point the velocity and temperature change becomes *quasi-periodic*. In these state the properties of the system change almost periodically. When ΔT exceeds $10.6 \ \Delta T_C$, the quasi-periodic behavior disappears and the fluid motion is completely random and *chaotic or turbulent*. The succession of these events is shown in Fig. 6.6, as was shown by G. Ahlers and R. Behringer.

The Benard system has the following characteristics: (1) it is a dissipative system which operates far from thermodynamic equilibrium; (2) it is endowed with self-organizational properties; (3), for discrete values of a parameter the various organized patterns appear in succession; and (4) the system experiences symmetry-breaking instabilities. A convection cell is a *dissipative structure* as defined by Prigogine, stressing the basic role of the viscous forces far from equilibrium. This also underlines the distinction that must be established from equilibrium structures, such as crystals.

6.6. TAYLOR INSTABILITY

In another example, a fluid is held between two concentric cylinders, as shown in Fig. 6.7a. The inner cylinder rotates around its axis and the external cylinder remains at rest. For small angular velocities (slow rotation) the fluid appears as uniform. But for a critical value of angular motion of the inner cylinder, various patterns appear in the fluid, as shown in Fig. 6.7b.

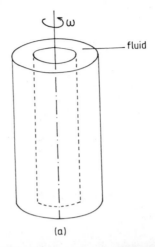

fluid

(a)

(b)

Fig. 6.7. (a) A viscous fluid is held between two cylinders. The inside cylinder rotates around a common axis. (b) For a critical value ω_C convection vortices appear and are made visible by addition of aluminum powder. (Courtesy of E. L. Koschmieder. From I. Prigogine and S. A. Rice, *Advances in Chemical Physics*, © 1975 Wiley. Reprinted by permission of John Wiley & Sons, Inc.)

Clearly the Taylor instability is another example of a dissipative structure. Patterns appear in the fluid whenever a parameter of the system exceeds a threshold value and the homogeneous steady state becomes unstable.

6.7. INSTABILITY IN A CHEMICAL REACTION

The two examples of the preceding sections, both taken from hydrodynamics, would have remained scientific curiosities were it not for the fact that they are part of a large class of physical and chemical systems exhibiting self-organizing properties.

In both examples the shapeless fluid spontaneously organizes itself into rolls or hexagons, or into superimposed fluid layers. The important fact is that it may do so if energy is dissipated by the system. This is only possible if the system is held in a nonequilibrium state by a constant input of energy. Moreover, we have seen that only the steady states sufficiently far from equilibrium can become unstable. Once the supply of energy is suppressed, the system returns to the original shapeless state. As we shall see, the self-organizing properties of a physical system are of primordial importance for an understanding of the emergence and functioning of biological systems.

Many biochemical reactions operate in the nonlinear, nonequilibrium domain. Thus there is some possibility that the steady states of these reactions will become unstable when a threshold value of reaction affinity is exceeded. One way to check this possibility theoretically is to construct model chemical schemes made of monomolecular, bimolecular, and trimolecular steps, and to investigate the contribution of each elementary step to the global stability of the chemical process.

We consider a model chemical reactor which essentially in an open chemical system, as depicted in Fig. 9.1. Reactant B enters and product C leaves the reactor at a constant rate and the medium is constantly stirred in order to keep the mixture uniform. With a proper choice of constraints the chemical reaction reaches a far from equilibrium steady state.

In the reactor, one mole of B combines with two moles of an intermediary substance X to produce three moles of the same substance. Such processes are called *autocatalytic*. X in turn transforms into the final product C. Both reaction steps in the Schlögl model are reversible:

$$B + 2X \underset{k_{-1}}{\overset{k_1}{\rightleftharpoons}} 3X$$

$$X \underset{k_{-2}}{\overset{k_2}{\rightleftharpoons}} C \tag{6.14}$$

At steady state the concentration of the intermediate substance assumes a time-independent value X^s. The stability of this state may by tested by introducing a small fluctuation into the reactor.

Following the methods of Section 6.4, we must evaluate the excess entropy production of the total scheme, Eq. (6.14), which is the sum of the excess entropy production of all elementary steps.

Provided B and C are held constant, the steady-state affinity of the first step, given by Eq. (3.40), is $A^s = RT \ln(k_1 B/k_{-1} X)$; while the velocity of the reaction is given by $v^s = k_1 BX^2 - k_{-1} X^3$. Under the influence of fluctuations the force and the flux take the new value

$$A_1 = RT \ln \frac{k_1 B}{k_{-1}(X^s + \delta X)}$$

$$v_1 = k_1 B(X^s + \delta X)^2 - k_{-1}(X^s + \delta X)^3 = v_1^s + \delta v_1 \qquad (6.15)$$

As δX is small, both expressions of Eq. (6.15) may be expanded in a Taylor series, and one immediately finds

$$\frac{\delta A_1}{T} = -R \frac{\delta X}{X^s}$$

$$\delta v_1 = (2k_1 BX^s - 3k_{-1} X^{s^2})\delta X$$

From Eq. (6.10) we see that

$$\delta_X P = \int dV \delta J_k \delta X_k = \int dV \delta_X \sigma$$

$$\delta_{1X}\sigma = \delta A_1 \delta v_1 = -R(2k_1 B - 3k_{-1} X^s)(\delta X)^2$$

If $3k_{-1} X^s < 2k_1 B$, as X and R are positive quantities, then $\delta_{1X}\sigma$ is negative, and therefore the autocatalytic step gives a destabilizing contribution to the total excess entropy production.

$\delta_{2X}\sigma$ may be calculated immediately from the second step, and is always positive,

$$\delta_{2X}\sigma = Rk_2 \frac{(\delta X)^2}{X^s}$$

The excess entropy production of the total system is therefore

$$\delta_X \sigma = R \left[\frac{k_2}{X^s} + 3k_{-1} X^s - 2k_1 B \right](\delta X)^2 \qquad (6.16)$$

In a certain range of values of the parameters, Eq. (6.16) may become negative, in which case the homogeneous steady state of the system cannot survive. Therefore the system is driven to a new and more stable state.

The reader may verify immediately that a bimolecular autocatalytic step $B + X \rightleftharpoons 2X$ may also give a negative $\delta_X \sigma$. Moreover, it can be shown that in the linear domain all autocatalytic steps are stable. Thus with this simple example we have shown that far from thermodynamic equilibrium autocatalytic chemical reactions may become unstable.

We shall see in the next chapter that real chemical reactions do exhibit such instabilities, and we may wonder what kind of behavior the system adopts after leaving the uniform steady state. From the examples of Benard and Taylor instabilities, we might expect a succession of ordered structures, and perhaps also temporal or spatio-temporal self-organization.

6.8. ORDER AND ENTROPY

Thermodynamic theory studies bulk matter without any explicit reference to the atomic nature of matter. However, in mid nineteenth century a parallel science to thermodynamics, given the name of Statistical Mechanics, was developed for the understanding of bulk matter, based this time on the atomic nature and motion of individual particles. Naturally, soon connections were established between these two branches of physics. In 1862, Boltzmann showed that the abstract and macroscopic concept of entropy could be related to a measure of molecular disorder. In this interpretation, the law of increase of entropy is equivalent to the progressive disorganization of the system. This is why, if we stop heating, the molecular cooperation which produces the rolls of Fig. 6.5 disappears, and the system reaches thermal equilibrium characterized by a maximum entropy.

In another example, a drop of ink is deposited on the surface of clear water. We thus obtain an organized, two-color pattern, as shown in Fig. 4.6. However, the forces generated by the differences in the chemical potentials of the two substances drive the system toward thermodynamic equilibrium, and the ink completely dissolves in water, erasing the initial ordered pattern. This is why today the word *entropy* is used even in nonscientific language to mean disorder.

Order may also be created in an equilibrium thermodynamic system if the system is acted upon by an external agent. For example, in the experiment shown in Fig. 6.1, if the thermostat is held at 100°C, all water transforms into vapor. In a sufficiently large container water molecules start moving almost freely and practically independently from each other. If we lower the temperature of the thermostat to 30°C, vapor may transform again into water.

In this state, molecules are much more organized and are correlated. Finally, if the thermostat reaches the freezing temperature of 0°C, water transforms into ice crystals. These are highly ordered, and strong correlations exist between molecules, which form a regular crystalline lattice.

For nearly two centuries, although thermodynamic theory was a well-established branch of macroscopic sciences, it was nevertheless confined mostly to the fields of physics and chemistry. Thermodynamics was useful in biological sciences whenever a given biochemical reaction could be considered as a purely equilibrium chemical process. But in general thermodynamics failed the biological sciences in many respects. The prime characteristics of biological systems are temporal and spatial order. Organisms are born, develop, reproduce, and die. All these events unfold in time in a well-defined order. A familiar example of a biological temporal order is the heartbeat in man and animals. In normal conditions the heart beats 70 times per minute, independently of whether the individual is in the heat of an equatorial forest or in the freezing temperatures of the north pole. But a heart stops beating if deprived of oxygen for only a few minutes. We were not able to keep a drop of ink over a glass of water, but nature has managed to keep stripes on the zebra and spots on the leopard. Biological organisms with a high content of water molecules function well above the temperature of formation of crystalline order. Moreover, biological order seems to be an intrinsic property of a given organism and its individual and specific response to external constraints, whereas equilibrium order is predictable from environmental conditions.

The equilibrium thermodynamics of the nineteenth century was based on the second law, which predicted a gradual disorganization of the system. It could not account for the daily observations which showed the reverse phenomena. Consequently, vitalistic theories were invoked whereby it was suggested that biological organisms obey laws that are not part of ordinary physics and chemistry. The development of nonequilibrium thermodynamics extended the domain of application of thermodynamic methods to biological systems—examples of membrane permeability and active transport were given in Section 4.11. However, as we shall see later, the variety and complexity of biological organization is such that a gradient in the distribution of matter generated by the maintenance of a temperature gradient of thermodiffusional nature is not appropriate for an understanding of biological order.

The discovery of the unstable nature of a class of steady states of matter which are far from thermodynamic equilibrium, made possible the understanding of the spontaneous onset of time and space organization in a given physico-chemical system. If a homogeneous steady state is unstable for a threshold value of a certain parameter, it must necessarily move to a nonhomogeneous, or time-dependent, or time- and space-dependent regime, or some other organized state. This switch to an organized state is an intrinsic property of the particular system under consideration, and is not the privilege of every thermodynamic system operating under the same constraints. Thus the stability criterion, Eq. (6.13), is a link between physico-chemical systems and biological organization. It is important to notice that with the methods of nonlinear, far-from-equilibrium thermodynamics, we may only predict the possibility or the impossibility of organized states in a given system. Thermodynamic methods cannot predict the nature of states available to the unstable homogeneous system. These new states may only be found in the framework of a dynamic approach to the study of matter.

Now it is time to ask which kind of open systems may spontaneously generate the spatial, temporal, or spatio-temporal order which is the rule in biological systems. From the previous chapters we might infer, and experience shows, that most transport processes may be studied quite satisfactorily by linear laws, such as Fick's law of diffusion, Eq. (4.20), or Fourier's law of heat conduction, Eq. (4.19). Thus transport phenomena alone are not sufficient for introducing organization in a nonequilibrium system.

However, we must note that whenever in a system the convection processes must be considered, as was the case in Benard convection, self-organization may appear. Nevertheless, such large-scale convection currents, initiated by temperature gradients, have not been detected in biological systems. On the other hand, we have seen that most chemical reactions operate generally far from thermodynamic equilibrium, and also that a uniform state of a reaction with autocatalytic steps may become unstable. In the case of reaction schemes with feedback mechanisms, that is, whenever a product enhances or decreases the production of its precursors, as shown in Eq. (6.14), such schemes may also exhibit instabilities of homogeneous steady states.

Biological systems are constructed from a long chain of many thousands of biochemical reactions. Feedback mechanisms are widespread and enzymatic catalysis is the rule. In addition, biological organisms are open systems that exchange matter and energy with their surroundings; thus they function as dissipative units. Therefore, it seems likely that biological order is related to the instability of far from equilibrium biochemical reactions that function via feedback mechanisms.

CONCLUSION TO PART I

We have seen that for two centuries the science of thermodynamics dominated and sealed the fate of industrial societies. With the help of its laws, heat engines were improved, automobiles were designed, and the chemical industry could develop. Therefore, it was natural to use these successful methods for the understanding of life and its origin. However, although chemical thermodynamics can be used for the study of biochemical processes, the all-important second law, endowed with the tendency to drive systems toward maximum disorder, prevented thermodynamic theory from accounting for the temporal and spatial order which characterizes living organisms.

The development by P. Glansdorff and I. Prigogine of evolutionary and stability criteria for nonequilibrium systems reconciled the living world with thermodynamic theory.

Once scientists were convinced that there is no contradiction between the laws of macroscopic physics and the self-organizational properties of matter and vital biological functions, new paths of research were opened for the investigation of these processes.

The self-organizational properties of matter are the subject of Part II. There we also show all the various options open to an unstable thermodynamic system.

Part II

A NEW CHEMISTRY

Order out of chaos

I. Prigogine 1983

INTRODUCTION TO PART II

The twentieth century has witnessed a turning point in the chemical and biochemical sciences. New chemical reactions were discovered exhibiting unexpected properties and requiring new concepts for their understanding; a *new chemistry* was born. The need to reconsider our view of some aspects of chemical reactions did not come from any breakthrough in our knowledge of molecules, atoms, or elementary particles. On the contrary, we are back again to considering bulk matter and its supramolecular properties, but this time in a completely new light.

In order to see, measure, and understand this new chemistry, we must do away with the idea of matter as something static and consider it in a dynamic framework. Matter and energy must flow continuously, chemical reactions must never stop, and thermodynamic equilibrium must be avoided at all cost. In such an environment, atoms and molecules cooperate, sometimes at the scale of the entire system, and produce bulk properties that could not be suspected from the behavior of the molecules and atoms.

Molecular cooperation introduces several types of organized supramolecular states in otherwise disorganized matter. These states appear spontaneously as the intrinsic properties of a given reaction and in the absence of any outside organizing factors. We thus speak of the *self-organization* of homogeneous matter. On the other hand, as this self-organization requires constant input and output of matter and energy, we also speak of *dissipative structures*.

The cooperative properties of molecules are most conveniently studied with the help of the mathematical tools mentioned in the introduction. The framework of differential equations is the most convenient formalism for a dynamic description of reacting media. However, as we shall see, because of the nonlinearity of chemical processes, the mathematics necessitated by the new chemistry is extremely complex, and one is led to solving nonlinear differential or partial differential equations for which general methods of solution are not available.

Part II of this book offers a broad outline of the various aspects of the new chemistry. In Chapter 7, we summarize the various self-organizing phenomena that were discovered in several different fields of science and gave birth to the concept of dissipative structures. Chapter 8 describes various modes of self-organization that have been observed experimentally in a real chemical reaction. Finally, Chapter 9 introduces mathematical tools for the study of self-organization and several model reactions which describe the observed experimental phenomena.

Chapter 7

THE BIRTH
OF THE NEW CHEMISTRY

7.1. INTRODUCTION

Since the end of the nineteenth century, a few isolated self-organized states of matter have been independently observed in chemical laboratories, chemical reactors, ecosystems, and in heated fluids. At the same time the observed patterns and ordered structures in living organisms had long been a puzzle to embryologists and were crying out for an explanation based on the physico-chemical laws of nature. In the first decades of the present century, the mathematical formalism for the description of self-organized states became available. This formalism was primarily developed for the study of planetary motion. The new progress in the area of nonequilibrium thermodynamics and the realization that self-organization and the second law of thermodynamics were not necessarily contradictory shed a new light on the possible relationships among the seemingly diverse fields mentioned above.

The new chemistry we are about to describe emerged from the convergence of various streams, all flowing from the so-called "hard" sciences. It is a multidisciplinary study. Once the concepts of the new chemistry were firmly established, they could be turned around, leading to cross-fertilization and illumination of the areas from which they emerged.

7.2. CHEMICAL OSCILLATIONS

It all started with strange, oscillatory chemical reactions. Following the tradition of Mesopotamian glassmakers and medieval alchemists, most twentieth century chemists still performed their reactions in closed systems. Various reactants were united in a vessel, forming a closed thermodynamic system, in such proportions as to keep the reaction initially very far from equilibrium. In such conditions the quantity of the new-formed products increases gradually until the mass action law, Eq. (2.14), is satisfied at equilibrium state. This is the obvious procedure for synthesis, which is what most chemists do.

Until the beginning of the present century, all chemical reactions seemed to

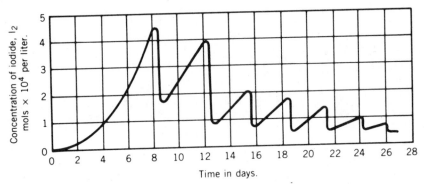

Fig. 7.1. The first oscillatory reaction involving IO_3^- and hydrogen peroxide was discovered by Bray in 1921. (Reprinted with permission from J. Amer. Chem. Soc. Copyright 1921 American Chemical Society.)

follow this regular pattern, predicted by the general laws of equilibrium thermodynamics. However, in 1916, Morgan was experimenting with a reaction medium that contained hydrogen peroxide (a common antiseptic and bleaching agent), formic acid, (a substance secreted by ants), and sulfuric acid; he observed a periodic release of deadly carbon monoxide from the reaction medium. Later, Bray found another periodic reaction. Again the acidic medium contained peroxide molecules, but this time iodate, IO_3^-, ions, were substituted for formic acid. Figure 7.1 shows the oscillatory behavior of the Bray reaction.

These few and "abnormal" oscillating chemical reactions went almost unnoticed by most chemists until the 1960s. These exceptional reactions were considered as objects of curiosity or as artifacts, since they were not predictable and did not appear to follow the chemical equilibrium laws which rule out any coherent behavior such as sustained oscillations in chemical concentrations.

7.3.   THERMOKINETIC OSCILLATIONS

Chemical engineers had been aware of oscillatory phenomena much earlier. In the chemical industry, large volume, that is, a continuous product output, is required; and chemical reactions are performed mostly in continuously stirred tank reactors, or CSTR. In these systems, products, at given concentrations $C^\circ(T^\circ)$ and temperature T° enter the reaction chamber with constant rates, and after a given residence time τ leave the reacting medium with temperature T and concentrations C_i. Figure 7.2 depicts such a CSTR reactor, which constitutes an open system, functioning at a steady state regime. However, the

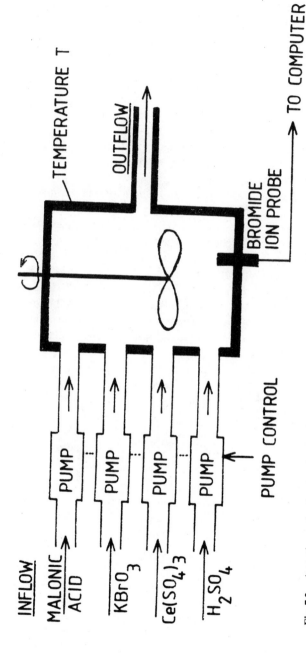

Fig. 7.2. A continuously stirred tank reactor, or CSTR. A nonequilibrium condition is maintained by regularly pumping reactants from the reservoirs into the reactor. A photoelectric cell monitors the periodic color change. (Courtesy of I. Prigogine.)

143

temperature of these steady states is not always stable, and time-periodic phenomena are generated, especially if the reaction mixture is in the gaseous phase and is heated or is subject to self-heating as a result of exothermic processes. These instabilities arise because of the fact that chemical reaction rates are related to the temperature via highly nonlinear functions. The temperature oscillations may range from a few degrees up to 100°C. Let us illustrate these phenomena with an example long known to chemical engineers.

Propane is a common combustion gas which is often used by campers. In open air it combines with oxygen and burns with a continuous flame. If the propane stove is properly adjusted, the temperature at every point of the burning flame may be kept constant in time. However, the same propane and oxygen may be burned, or rather *oxidized*, in an incomplete combustion inside a CSTR. In this case equimolar quantities of propane and oxygen which have been preheated separately are introduced into the reactor at a constant rate. If the temperature of the oven does not exceed 310°C and the residence time is of the order of 10 seconds, the combustion is slow, the system reaches a steady state, and only a small temperature increase is observed in the CSTR. However in the narrow temperature range of $T = 311$ to 313°C, at steady state, the temperature profile exhibits striking oscillatory behavior, as shown in Fig. 7.3.

The oscillatory nature of such processes became apparent in the late 1930s, and soon Frank-Kamenetskii postulated that they were oscillatory reactions. Interestingly enough, in a first attempt to explain the oscillatory behavior, he borrowed a mathematical formalism from population dynamics. The Volterra-Lotka mechanism that he used was originally developed for describing the interdependence of animal species.

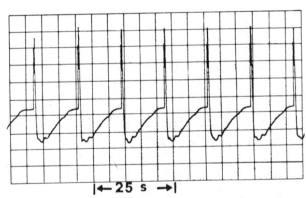

|← 25 s →|

Fig. 7.3. Periodic two-stage ignition. Each large temperature pulse is accompanied by a large pressure pulse and bright light emission. (From P. G. Lignola et al., 1980, reprinted by permission of VCH Verlagsgesellschaft.)

7.4. OSCILLATIONS IN PREY-PREDATOR POPULATIONS

At the beginning of the twentieth century, mathematical methods were widely used for describing the growth of bacterial populations and the progress of various diseases, such as malaria. These problems were discussed in terms of differential equations in which time is the variable and the functions are the average values of the population densities. The relevant differential equations were constructed by observing the behavior of nature.

Most living species are dependent on one, if not many, other species for their survival. For example, regular and periodic changes in the population of a parasite species and its host were observed by entomologists in the nineteenth century. In 1920, Lotka put forward a mathematical description of such periodic behavior for the general problem of coexistence of prey and predator populations. For example, we consider two neighboring population, those of the lynx and the snowshoe hare. The latter is a vegetarian, the former a carnivore who hunts snowshoe hares. This crucial survival association must be regulated in such a manner that the lynxes must not eat all the snowshoe hares in their environment, or their own species will disappear.

One solution to this problem may be that the ratio of prey and predator populations remains always constant. Another alternative is that in the presence of a large prey population, a small predator population increases beyond the level that can be supported by the prey population. As the number of prey decreases, in the absence of sufficient food the predator population also decreases. As the predators disappear, the prey population may increase again, generating in turn a new increase in the predator population, and a new cycle begins. Figure 7.4 shows such an oscillation in the prey-predator populations of lynx and snowshoe hares.

In the Volterra-Lotka model N_1 and N_2 are respectively the numbers of the prey and predator populations. The change in time of these numbers is given by

$$\frac{dN_1}{dt} = k_1 A N_1 - s N_2 N_1$$

$$\frac{dN_2}{dt} = s N_1 N_2 - k_2 N_2 \tag{7.1}$$

$N_1(t)$ and $N_2(t)$, the time-dependent solutions of these coupled nonlinear differential equations, must therefore exhibit a periodic behavior. As we shall see in Chapter 9, Eq. (7.1) do indeed have oscillatory solutions. However, since the period of oscillations is a function of the initial prey and predator populations, there are many different modes of oscillation.

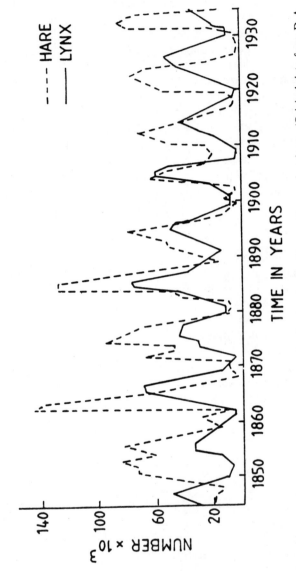

Fig. 7.4. Prey-predator oscillation. Changes in the abundance of lynx and snowshoe hares. (Original data from D. A. McLulich, 1937.)

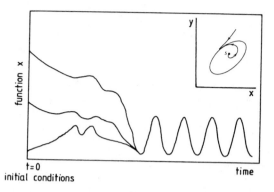

Fig. 7.5. A limit cycle. Two coupled differential equations $dx/dt = f(x, y)$, $dy/dt = g(x, y)$ occasionally admit solutions which, for every initial conditions, if represented in the x-y plane, are closed curves called limit cycles.

7.5. A MATHEMATICAL APPROACH TO OSCILLATORY BEHAVIOR

In 1928 Poincaré, and later Andronov and his school, approached the problem of nonlinear differential equations from the mathematical end. They showed that a large class of two coupled nonlinear differential equations may exhibit an oscillatory solution of a very special kind. Contrary to the Lotka-Volterra equations, the final behavior of these equations was independent of the initial conditions. No matter what the initial state of the system, it invariably reaches the same periodic motion, called a *limit cycle*. Figure 7.5 shows such a time-periodic solution, which only arises in certain nonlinear coupled differential equations. The existence of such periodic motions are of crucial importance for the regulatory processes of biological organisms, which can only be a function of various parameters of the system and are completely independent of any initial conditions.

7.6. PERIODIC BEHAVIOR AND SPACE ORDER IN LIVING ORGANISMS

Living organisms are characterized by functional and morphological order. Seasonal periodicities of plants and various spatial organizations of niche structures may be observed in nature. The most familiar functional order is seen in the heartbeat of animals, while another example is found in the *circadian rhythms*. These are the various periodic phenomena which appear with a period of approximately one day. These temporal organizations are endogenous; the rhythm continues even if the subject is isolated from diurnal influence. For example, *Kalanchoe biossfeldiana* is a tiny red flower that normally opens and closes every 23 hours. If the flower is cut from the plant and placed in a sugar solution, at constant temperature and illuminated with constant green light, it continues to follow its natural circadian rhythm for up to a week.

Many examples of rhythmic behavior may be found in nature. Large-scale biological oscillations are the rule rather than the exception. Such oscillations require the cooperation of a great many cells. In the 1960s, a few oscillatory biochemical processes have been discovered that operate at the cellular level. For example, the synthesis of some proteins by cells follows an oscillatory pattern. Another example is furnished by the sugar consumption of yeast cells, which necessitates several transformations involving energetically important biomolecules. Concentrations of some of the intermediates of this process change periodically in time.

Morphological order is the most conspicuous attribute of living species. Roots, leaves, branches, skin, hair, legs, all appear with a well-defined order and relationship with respect to each other. Moreover, this morphological order and its sequential unfolding in space and time is engraved in the tiny seeds and minute fertilized eggs. Figure 7.6 shows an example of a morphological order.

Understanding biochemical events with the help of the same laws which describe inanimate nature has been a longstanding challenge for embryologists. At the end of the nineteenth century Drisch proposed that *morphogenesis*, the birth of forms, is the consequence of the onset of various

Fig. 7.6. A morphological order.

gradients of undetermined nature in the developing embryo. In order to substantiate these ideas, in 1952 Turing assumed that some biochemical reactions inside the developing organism may be of a catalytic or cross-catalytic nature, and, moreover that the various substances may diffuse throughout the organism. He could then show that correlations between chemical reactions and transport processes may spontaneously generate chemical concentration patterns in the embryo. Thus a physico-chemical basis for Drisch's hypothesis was found. Turing was a mathematician and his morphogenetic patterns stemmed from the solution of a set of two coupled nonlinear partial differential equations of the type $\partial C_i/\partial t = f_i(C_1, C_2) + D(\partial^2 C_i/\partial r^2)$. These equations were constructed from the juxtaposition of a kinetic part, $f_i(C_1, C_2)$ and a contribution from diffusion processes given by $D(\partial^2 C_i/\partial r^2)$. Unfortunately, Turing's work went unnoticed by the embryologists and chemists of his time.

7.7. THE CONTRIBUTION FROM THERMODYNAMICS

We have stated that the equilibrium thermodynamics of closed systems seemed inadequate for the description of temporal periodicities and spatial order in biological systems. In the 1940s, in order to reconcile thermodynamics with biological organization, Bertalanffy, Schrödinger, and Prigogine drew attention to the fact that biological organisms are open systems. In fact, a cell of any living organism only survives if it continuously exchanges matter with its surroundings. Therefore, living structures need not necessarily follow the law of increasing disorder. On the other hand, in order to perform vital tasks cells must generate a nonequilibrium condition in their environment. The example of active transport has already been given in Chapter 4. If the reader is able to decode these words, it is because of the fact that a nonequilibrium distribution of Na^+ and K^+ ions, at a very high energetic cost, takes place across the membranes of his brain cells.

The development by Glansdorff and Prigogine of the thermodynamics of far from equilibrium processes was a great leap forward in our understanding of biological systems. The stability criterion, Eq. (6.13), showed that thermodynamic methods can predict the onset of spatio-temporal order in an open chemical system. Moreover, such systems must necessarily evolve according to nonlinear kinetics, and therefore they must be described by nonlinear coupled differential equations of the type that produces limit cycles. Biochemical mechanisms also follow in most cases nonlinear kinetics. Therefore, it is likely that a great number of vital processes in living organisms could produce the observed spatio-temporal order in complete accord with the macroscopic laws of nonequilibrium thermodynamics.

7.8. ORGANIZED PATTERNS IN FLUIDS

To this long list of organized behavior we must add the Rayleigh-Benard instabilities already discussed in Part I. Let us recall that if a layer of a fluid is held between a temperature gradient, for a well-defined value of the latter, the fluid does not remain homogeneous, and hexagonal patterns arise. These patterns result from the cooperation of a great number of molecules; one example is shown in Fig. 7.7. Organized states in fluids had been observed by the end of nineteenth century, and a theoretical explanation was derived by Rayleigh in 1916. However, only recently have family ties between hydrodynamic and chemical self-organization become apparent. Moreover, it was shown that these two phenomena are special cases of *dissipative structures* as defined in the next section.

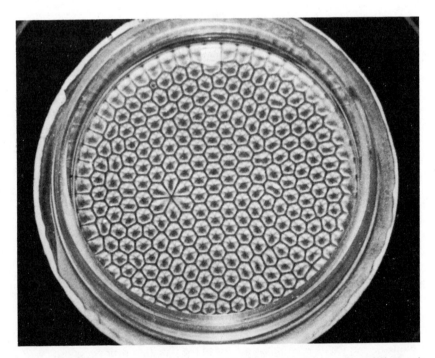

Fig. 7.7. Self-organization in fluids. Silicone oil is heated uniformly from below. Under an air surface, hexagonal Benard convection cells are seen in the presence of aluminum powder. (Courtesy of E. L. Koschmieder. From *Order and Fluctuations in Equilibrium and Nonequilibrium Statistical Mechanics*, G. Nicolis, G. Dewel, and J. W. Turner. © 1981. Reprinted by permission of John Wiley & Sons, Inc.)

7.9. THE CONCEPT OF DISSIPATIVE STRUCTURES

With the development of nonequilibrium thermodynamics as a conceptual framework, Prigogine and his colleagues could show that all these seemingly diverse temporal and spatial orders could be unified into the single concept of dissipative structures (see Fig. 7.8). Dissipative structures include all periodic behavior, spatial order, spatio-temporal regularities, or any other coherent property seen in a system as a consequence of the instability of the homogeneous state of that system. More specifically, the conditions for the onset of dissipative structures are:

1. The system must be open and subject to a constant input and output of matter and energy. This fact implies that chemical, biochemical, and

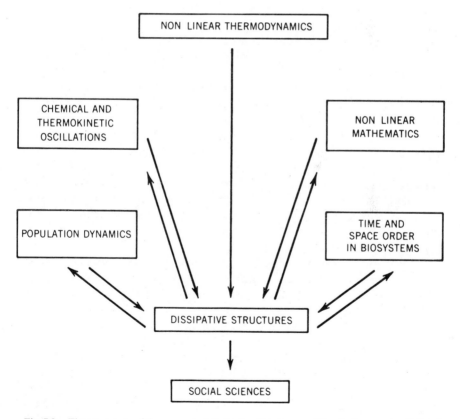

Fig. 7.8. The emergence of the concept of dissipative structures and its relation to other fields of science.

hydrodynamic systems must be kept far from thermodynamic equilibrium.

2. Various catalytic, cross-catalytic, or feedback processes must be present in the system. This fact insures the description of the system in terms of non-linear differential equations.

3. For well-defined values of imposed constraints, the homogeneous steady state can no longer damp the small fluctuations present in all environments, and a new organized and stable state appears in the system.

The most important characteristic of dissipative structures is that they only appear and are maintained as long as nonequilibrium conditions are applied. Under the influence of these constraints, the units of the global system cooperate and create a behavior that is the property of the system as a whole and cannot be understood or suspected from the study of the properties of the individual members. For example, in chemical systems as well as in Benard and Taylor instabilities, many molecules cooperate in order to create temporal and spatial order. Thus dissipative structures appear whenever a system is able to *self-organize* itself in a *cooperative* fashion, measure time, and organize space in order to "survive" under the imposed constraints, or to better exploit the environment.

In a second stage, the concept of dissipative structures catalyzed and triggered interest in those areas which gave birth to it. A revival of interest arose in the field of nonlinear differential equations. Mathematical models for the description of various biological processes in single cells or in multicellular ensembles became very popular. The chemistry of oscillating reactions was taken up, and soon many other unexpected self-organizing behaviors were discovered in chemical reactors (see Chapter 8). In turn, hydrodynamic instabilities again became the focus of extensive research. New and unexpected organized, quasi-periodic, and chaotic behaviors were found in fluids. From the convergence of all these sources a new area of research, called *nonlinear phenomena*, has emerged and is currently one of the most promising fields of macroscopic physics.

Let us stress again that the most important notion embedded in the concept of dissipative structures is the possibility of self-organization occurring only far from thermodynamic equilibrium. The necessity for investigating the properties of physico-chemical systems far from thermodynamic equilibrium came primarily from the desire to understand the functioning of living organisms which obviously are open systems. Once methods for the study of far from equilibrium organized states became available, they invaded the fields of biology, physics, and chemistry. A fruitful cross-fertilization occurred which gave rise to a new chemistry, revived nonlinear macroscopic physics, and gave birth to a dynamic view of biochemistry.

When pertinent variables of a dynamic system change with time, they may usually be described by differential equations. If competition, self-enhancement, or self-destruction of entities are possible, then the differential equations are nonlinear. Formally these equations may become identical to the differential equations underlying chemical dissipative structures. Thus one can expect that such systems should exhibit cooperative behavior and self-organize themselves spontaneously into various modes of spatio-temporal behavior. This is why in recent years the framework of dissipative structures has been extended to such fields as sociobiology, sociology, socio-economics, and economics.

*　　*

*

In the next chapter we study in some detail the Belousov-Zhabotinski reaction, which is one of the best-known chemical clocks. This reaction exhibits other unexpected self-organizing behaviors which may be seen as prototypes for various dissipative structures encountered in living organisms.

Chapter 8

THE BELOUSOV-
ZHABOTINSKI REACTION

8.1. A CHEMICAL CLOCK

The Belousov-Zhabotinski reaction is one of the best-studied oscillatory chemical processes. It alone exhibits most properties of dissipative structures predicted from theoretical considerations. In 1958 Belousov, a Russian chemist, was studying an intermediary step of a complex chemical mechanism. He was led to investigate the oxidation, the loss of electrons by molecules, of citric acid by potassium bromate, $KBrO_3$. The reaction proceeded in acid solution (a medium containing more H^+ ions than water) in the presence of cerium ions, Ce^{3+}, which served double duty as a catalyst (they were not consumed by the reaction) and as an indicator dye. To his great surprise, the homogeneous solution oscillated with clockwise precision between a pale yellow and a colorless state, as shown in Fig. 8.1.

The reaction medium contained cerium ions in two different ionic states. The Ce^{4+} ions are pale yellow, while the Ce^{3+} ions are colorless. The measurements of concentration change in time of the ratio $[Ce^{4+}]/[Ce^{3+}]$ exhibited regular time-periodic behavior, as shown in Fig. 8.2, thus providing a quantitative confirmation of the oscillatory behavior. A few years later, Zhabotinski repeated and confirmed Belousov's experiments, substituting for citric acid another organic compound, malonic acid. The news of this amazing reaction reached the western world, and several laboratories in the Soviet Union, the United States, and Western Europe started studying what came to be known as the Belousov-Zhabotinski reaction. Oscillating reactions had at

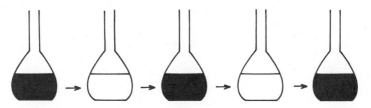

Fig. 8.1. A chemical clock. The reacting mixture oscillates periodically between colorless and yellow state.

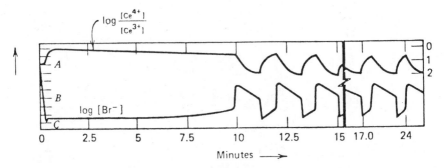

Fig. 8.2. Potentiometric traces at 25° of log [Br⁻] and log [Ce⁴⁺]/[Ce³⁺] versus time during Belousov reaction. Initial concentrations: $CH_2(COOH)_2 = 0.032 M$, $NaBrO_3 = 0.063 M$, $Ce(NH_4)_2(NO_3)_5 = 0.01 M$, $H_2SO_4 = 0.8 M$ and $KBr = 1.5 \times 10^{-5} M$. (From R. J. Field, 1972, reprinted by permission *J. Chem. Educ.*)

last entered the chemical laboratories as a respectable and "hot" subject of research.

As investigations progressed, much to the delight and amazement of the scientific community other extraordinary properties were found in Belousov-Zhabotinski (henceforth BZ) mixtures with appropriate concentrations. The reaction mixture had also suffered a few minor alterations. The change in the reaction mood was made more dramatic by substitution of iron ions for cerium ions. The couple Fe^{2+}/Fe^{3+} can do exactly the same job as Ce^{3+}/Ce^{4+}, exhibiting alternate red and blue coloring when a certain organic dye is present. This makes the oscillations easier to observe, but does not change the basic phenomena.

Thermokinetic oscillations had earlier been observed by chemical engineers in open reactors; the phenomenon was the consequence of coupling between chemistry and thermal effects, whereas the BZ reaction was thought to be of purely chemical nature. Thus, chemists finally became convinced of the reality of chemical clocks. Besides the few old oscillatory reactions mentioned in the introduction, many new chemical clocks have been found in recent years. Their number is growing constantly. Figure 8.3 shows a few of the presently known oscillators.

8.2. HOW THE CHEMICAL CLOCK FUNCTIONS

Four reactants in an aqueous medium are the necessary ingredients of a BZ reaction. The global stoichiometric equation is

Sulfuric acid + malonic acid + cerium ammonium nitrate + sodium bromate = carbon dioxide + dibromoacetic acid + bromomalonic acid + water (8.1)

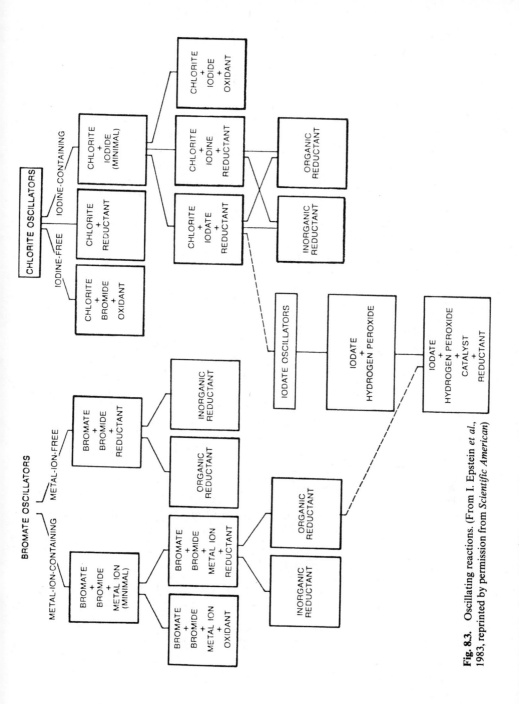

Fig. 8.3. Oscillating reactions. (From I. Epstein *et al.*, 1983, reprinted by permission from *Scientific American*)

157

This global equation hides some twenty or more elementary steps, many of which have not yet been completely elucidated. The reaction can be seen to be the oxidation (burning) of organic substances by bromate in acidic solution, in the presence of a catalyst, to produce carbon dioxide. Other organic acids can be substituted for malonic acid; iodine and chlorine derivatives can also have the same action as bromine compounds; the BZ reaction may be brought about in a variety of ways. A closer scrutiny of the reaction mechanism reveals the following processes. In aqueous solution sulfuric acid dissociates into two oppositely charged ions, HSO_4^- and H^+. Cerium nitrate dissociates into Ce^{4+} and a negatively charged ion, whereas sodium bromate furnishes Na^+ and BrO_3^- ions. The relevant components of the global reaction therefore are

$$2BrO_3^- + 3CH_2(COOH)_2 + 2H^+ \rightarrow 2BrCH(COOH)_2 + 3CO_2 + 4H_2O \tag{8.2}$$

$CH_2(COOH)_2$ and $BrCH(COOH)_2$ are, respectively, malonic and bromo-malonic acid. Cerium ions play the role of a catalyst, and are not consumed in the reaction process. This is why they do not appear in the global stoichiometric equation (8.2).

A deeper understanding of the various extraordinary phenomena we are about to discuss requires a knowledge of the elementary steps of the reaction mechanism. A close scrutiny of the known elementary steps reveals that they may be separated into three categories of reactions, as shown in Fig. 8.4. Any solution of sodium bromate always contains small quantities of reaction initiator Br^- as an impurity. In the first group of reactions, Br^-, BrO_3^-, and malonic acid, MA, are consumed and H_2O and bromomalonic acid, BMA, are produced. These slow reactions do not require the presence of cerium ions, and $HBrO_2$ is one of the important intermediates of this group of reactions. Once $HBrO_2$ molecules are present in the reaction mixture a second and more efficient pathway for production of BMA becomes possible. In this pathway BrO_3^-, MA, $HBrO_2$, and Ce^{3+} finally produce Ce^{4+}, H_2O, and BMA via several elementary steps. The third category of reactions comprises steps which use BMA, MA, and Ce^{4+} ions in order to produce CO_2 and Br^-, and to regenerate Ce^{3+}, thus beginning a new cycle. It should be noted that many of the intermediate substances are shared by all three groups of reactions.

The pathways of the first and second boxes of Fig. 8.4 are two competing mechanisms which use the same intermediate, $HBrO_2$. In the presence of a high concentration of Br^-, $HBrO_2$ is consumed in the first box. Initially, cerium is in the Ce^{3+} state and the mixture remains colorless. As the reaction proceeds, the concentration of Br^- decreases slowly and some $HBrO_2$ becomes available for the initiation of the second pathway. A close look at the various steps of the second box shows that one mole of $HBrO_2$ produces two moles of

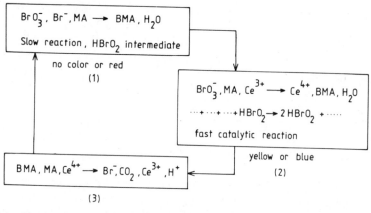

Fig. 8.4. The key intermediates of the BZ reaction. Some twenty elementary steps are needed for the description of the BZ reaction. In order to explicate the oscillating behavior, they are classified into three stages.

the same substance. Therefore the reaction is *autocatalytic* and evolves rapidly. All Ce^{3+} ions are transformed into Ce^{4+} and the reaction medium suddenly turns yellow (or blue).

Once Ce^{4+} ions are produced in the reaction mixture the mechanism of the third box is triggered, thus liberating Ce^{3+} and Br^-. The reacting medium again becomes colorless (or red). The newly formed Br^- may reach a critical concentration and switch the reaction course to the first pathway, and a new cycle begins. As long as BrO_3^- and malonic acid are in excess of their equilibrium value in the reaction vessel, they will be consumed by the two alternative paths, turning the reaction medium into a chemical clock which swings between colorless and yellow states with clocklike precision (as long as all the constraints are kept constant). In the experiment of Fig. 8.1 the oscillations continue as long as the initial products are in excess of their equilibrium values as given by the mass action law, Eq. (2.14). The frequency of the chemical clock or the number of times it changes color per unit time is a function of temperature and the initial concentration of the various constituents in the reactor.

However, the mixing of the ingredients of the BZ reaction as described in Eq. (8.1) does not necessarily ensure the onset of oscillatory behavior. The chemical clock only functions if one follows a well-defined quantitative recipe. This is also the case for all other types of spatio-temporal patterns that will be encountered later on. Each pattern arises only in a well-defined and rather narrow range of concentrations.

The chemical clock of Fig. 8.1 may last some thirty minutes or more; then it stops because of "energy" deficiency. However, it is possible to keep the clock oscillating for a long time, in principle indefinitely, if the reaction is performed in CSTR as in Fig. 7.2. In such experiments the reagents of the BZ reaction are continuously pumped into the reactor R. They enter R with a well-defined, adjustable, and constant rate. Matter also leaves the reactor by the overflow pipe P. The slower is the rate of entry of the reagents, the longer they remain in the reacting medium. We speak of long *residence times*. The inverse residence time thus is a measure of the distance from the equilibrium state.

8.3. PROPAGATION OF CHEMICAL INFORMATION

If the BZ oscillating reaction is performed in a petri dish containing a shallow layer of solution, patterns such as those of Fig. 8.5 appear gradually and fill the entire dish. In the presence of Fe^{2+}, Fe^{3+}, and an appropriate organic dye these two-dimensional patterns are formed by narrow circular bands of sharp blue leading edges and diffuse trailing edges that propagate through the motionless red mixture in regularly spaced successions. Their velocities are of the order of millimeters per second. The bands are generated at one or several centers, called *pacemakers*, either by dust particles or under the action of small local concentration fluctuations. Except for the first wave, which runs faster than the subsequent bands, the speed of propagation is approximately constant. Thus the following bands never overtake their predecessors, though they may collide with bands from other pacemakers and annihilate one another. When bands reach the boundaries of the vessel, they disappear without bouncing back again. These characteristics distinguish them from the familiar waves encountered in the oceans.

Usually several pacemakers signal simultaneously, each operating in a different area of the fluid. The periods of oscillation are variable, and the short-wavelength, high-frequency waves grow at the expense of the long-wavelength patterns. Under ideal conditions, the pacemaker with highest frequency eventually takes over and synchronizes the entire fluid.

The mechanism of the BZ reaction as shown in Fig. 8.4 can furnish a qualitative explanation for the onset of these bands, or *chemical waves*. The red motionless solution follows path 1 of the Fig. 8.4. Because of inhomogeneity in concentrations or temperature, locally some small portions of the reacting mixture suddenly may follow path 2, thus generating blue pacemakers. The reaction mixture is subject to interdiffusion of all species. The blue pacemaker region is rich in $HBrO_2$ which diffuses radially in all directions into the red solution. Owing to an efficient autocatalytic production, $HBrO_2$ switches the reacting medium, reached by diffusion, to path 2. We thus see a sharp blue front moving radially outward from the pacemaker region.

Behind the leading edge of a band, Ce^{4+} (or Fe^{3+}) accumulates, and the

Fig. 8.5. Randomly scattered pacemaker centers periodically generate circular waves. (Courtesy A. Pacault and C. Vidal.)

reactions of the third box are triggered, liberating Br^- and Ce^{3+} (or Fe^{2+}). When the level of Br^- ions rises sufficiently, the reaction switches to the red pathway 1. The color changes from blue to red in the diffuse trailing edge of the band.

The onset of the second and subsequent bands are due to the fact that the pacemaker can generate new bands with regular spacing. In a finer analysis, one must distinguish between oscillatory and excitable media, which have different mechanisms. Figure 8.6 shows the approximate concentration profile of the three key intermediates, Br^-, $HBrO_2$, and Ce^{4+}, in a typical train of bands moving from left to right.

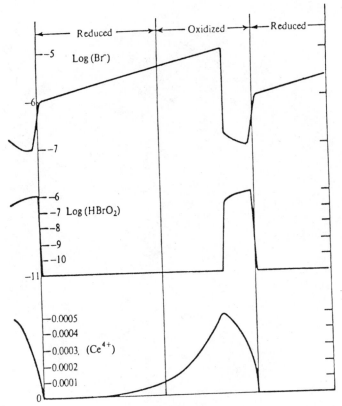

Fig. 8.6. Approximate concentration profile of the three key intermediates $[Br^-]$, $[HBrO_2]$ (logarithmic scales), and $[Ce^{4+}]$ (linear scale) in a train of bands moving from left to right. (From R. J. Field and R. M. Noyes, 1972, reprinted by permission from *Nature*, Vol. 237, p. 390. Copyright © 1972 Macmillon Journals Limited.)

These circular chemical waves may be visualized by an analogy with an example taken from the field of epidemiology. Imagine an extremely contagious flu virus that provokes an illness of a few days. Moreover, cured patients develop a short-term immunity to the disease. The virus starts to proliferate in a given geographical spot. It soon infects the population of the immediate surroundings. But people infected with the virus travel and bring the infection into new environments, where again a rapid extension of the illness is seen. Thus a wave front of flu propagates from the original center, provoking sometimes a worldwide epidemic. The wave-front structure is due to the fact that, behind the advancing wave, patients recover rapidly and are immune for a short time. If the infection center remains active or if the disease has not completely disappeared from an already infected area, a new wave front may start and propagate through the same regions.

The propagating chemical waves in an oscillating BZ reaction were first observed by Zaikin and Zhabotinski. Later, Winfree discovered that he could generate chemical waves in a BZ reaction of such a composition that the spontaneous homogeneous oscillations are excluded. This so-called *excitable medium* is perfectly stable until it experiences a perturbation of a magnitude exceeding a threshold value. For example, one may heat a small area locally (remember heat increases the local frequency of oscillations) until a blue state is initiated. It propagates through the medium just as if the reaction was in an oscillating state. These so-called trigger waves must be distinguished from the spatio-temporal patterns that we are about to describe in the following section.

8.4. STRIPES IN A TEST TUBE

The BZ reaction exhibits another unexpected and spectacular behavior which was discovered by Busse and may be demonstrated with rudimentary laboratory equipment. All we need is a test tube filled with a BZ mixture of a well-defined composition of all the reagents but one, for example the sulfuric acid. Following the addition of H_2SO_4 in the unstirred medium, colored stripes appear in the bottom of the tube and move upward. As time goes on, more and more bands appear and they move more and more slowly through the fluid, which is not itself in motion. The final pattern seems to be stationary and is formed from successive horizontal red and blue bands, as shown in Fig. 8.7. The patterns remain visible for some time, and finally vanish with the disappearance of nonequilibrium constraints in the tube.

The onset of these patterns is again the consequence of the oscillatory nature of the BZ reaction, and may be understood in the following manner. Sulfuric acid is heavier than the fluid in the test tube. Therefore, when added last, it tends to sink, thus creating a concentration gradient in the medium. We have seen that the frequency of the oscillations of the BZ reaction is very sensitive to the initial concentrations and temperature. In particular, the addition of sulfuric acid tends to increase the frequency of oscillations. Therefore, the solution in the bottom of the tube oscillates with a higher frequency than the liquid on top. A frequency gradient is established along the tube and the solution at each level is oscillating with its own specific frequency. Thus, superimposed fluid layers behave essentially like independent oscillators. The diffusion process as a homogenizing agent present in every system does not play any role in the present situation, since its action is too slow compared with the lifetime of the organized phenomena observed in a closed test tube.

The perception of band motion and the origin of the final stripes may be best

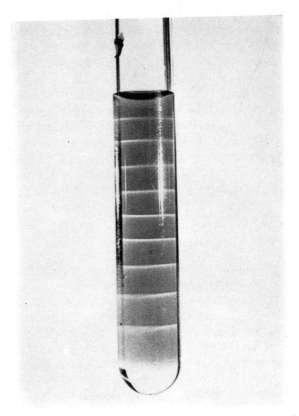

Fig. 8.7. Horizontal bands in the Belousov-Zhabotinski (BZ) reaction.

understood from the following experiment, taken from Pacault and Vidal. The experimental apparatus consists of a tank attached to a rotating device, as shown in Fig. 8.8. The tank is full of BZ mixture with concentrations picked to oscillate, say, every θ seconds. This period of oscillation, and therefore the frequency of the onset of the blue color, may be increased by the addition of various quantities of sulfuric acid. Ten adjacent rectangular vessels are part of the apparatus. Each nth vessel receives $2(10 - n)$ drops of concentrated sulfuric acid. The solution in the tank is made homogeneous by vigorous stirring, and is poured into the vessels by the rotation of the tank. Thus the solution in each vessel has a different composition; the period, θ_n of the chemical reaction is a linear function of n and is given by $\theta_n = \theta_1(n - 1)q$. Here q is the difference in the period between two adjacent cells. Immediately after pouring, at $t = 0$, the solutions in all vessels start to oscillate in unison or, as we say, *in phase*. However, the blue color appears for the first time in each vessel at time $t = \theta_n$, at different moments, thus giving the illusion of a propagating wave.

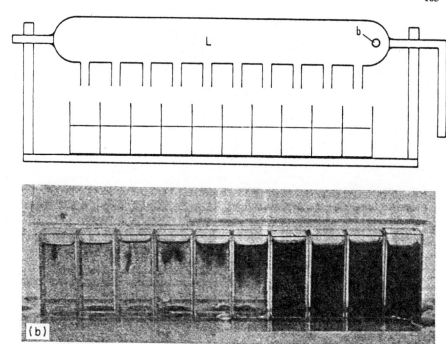

Fig. 8.8. (*a*) Experimental apparatus for the visualization of pseudo-waves. (*b*) An instantaneous view of pseudo waves. (From A. Pacault and C. Vidal, 1982, reprinted by permission from *J. Chimie Physique*.)

8.5. AMBIDEXTROUS CHEMISTRY

Let us consider again a BZ excitable or oscillating mixture, contained in a petri dish, which may sustain propagating circular chemical waves. If one of these propagating fronts is sheared by a silver pin, the two broken ends bend around and become foci of spiral patterns rotating in opposite directions and a pair of *spiral waves* propagate into the medium. The pitch of the spiral, and thus the frequency of rotation, is identical for the two sister spirals. Various pairs of spirals eventually fill the entire reaction space. However, the spirals do not have the tendency to grow at the expense of their neighbors. Such patterns are shown in Fig. 8.9. Agladze and Krinsky have managed to construct multi-armed spirals, as shown in Fig. 8.10.

The spiral formation may be understood with the help of the following qualitative arguments. Consider a pacemaker at point c of Fig. 8.11 and twelve points arranged in a circle around c. At every point there is a chemical clock

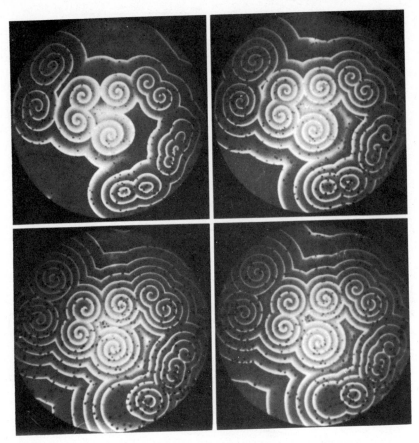

Fig. 8.9. A shallow layer of BZ mixture, if deformed by gentle stirring, produces spiral waves. The four photos show the time evolution of the spirals. Note that the two spirals rotate in opposite directions. (From A. T. Winfree, 1978, with permission of author.)

that changes into blue color every 60 seconds and remains so for, say, 5 seconds. All twelve oscillators are identical; however, they are *out of phase* by 5 seconds. If the first clock shows 1 P.M. the next clock will show this same time 5 seconds later, the third one 10 seconds later, and so on. If each oscillator is required to turn blue at 1 P.M. local time, then the blue color will appear successively in all twelve oscillators, thus giving the impression of a circular motion of a blue ribbon. However, the oscillating points are not static. The blue color at every point moves outwardly by the same mechanism as if it was

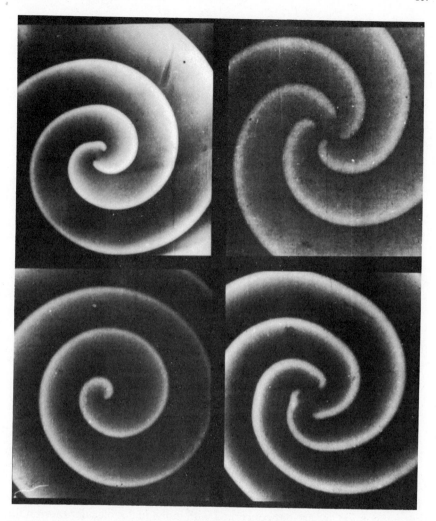

Fig. 8.10. One-, two-, three-, and four-armed spirals may be produced in an excitable BZ medium. (From K. I. Agladze and V. I. Krinsky, 1982, reprinted by permission from *Nature*, Vol. 29, pp. 424–426. Copyright © 1982 Macmillan Journals Limited.)

part of a circular concentric wave (see Section 8.3). The combination of these two processes produces the outwardly winding spiral wave propagation.

The important fact that the two sister spirals rotate in opposite directions introduces the concept of right- and left-handedness in an otherwise homogeneous and amorphous chemical medium. Moreover, if we concentrate on

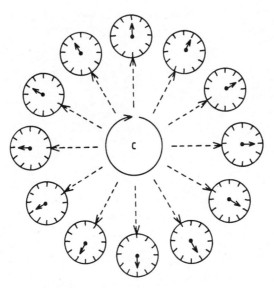

Fig. 8.11. Visualization of spiral propagation.

the region of fluid where only one pair of spirals exist, the space is no longer symmetrical and we speak of *symmetry-breaking instabilities*.

8.6. ERRATIC BEHAVIOR

Matter seen at the molecular level usually reveals a chaotic and disordered behavior. In a seemingly inert gas or liquid, molecules move in all directions in a chaotic motion. For example, a small drop of ink introduced into a cup of still water diffuses slowly into the liquid, thus revealing the random motion of molecules. However, when speaking of chaotic or turbulent behavior in the current context, we are not concerned with individual molecular motion but rather with the erratic behavior of bulk matter. Such unexpected chaotic behavior was recently found by Hudson in the BZ reaction performed in a CSTR reactor. The input and output fluxes are adjusted so that a stable sustained oscillatory regime prevails in the reactor. The concentrations of Br^- or Ce^{4+} are monitored and recorded in the course of time. Figure 8.12 shows such a recording.

If one of the flow rates across the reactor is decreased beyond a threshold value, the dynamic behavior of the system is radically modified. Time oscillations of concentration of Br^- are no longer regular. There are no longer well-defined oscillation frequencies. Such variations are known as *chemical*

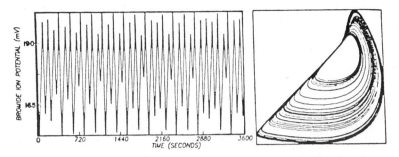

Fig. 8.12. If the BZ reaction is performed in a CSTR, under certain conditions time dependence of the bromide ion potential is irregular and is called chaotic. The second figure shows a two-dimensional phase portrait of a chaotic regime. (From J. S. Turner *et al.*, 1981, reprinted by permission *Phys. Lett.*)

turbulence or chaos, and are shown in Fig. 8.12. Surprisingly enough, if one decreases the flow rate still further, a simple periodic motion is seen again.

Note, again, that the chaotic regime just described is a property of the average macroscopic concentrations, and is not a reflection of the random fluctuations ever present in the system.

8.7. A CHEMICAL REACTION WITH MEMORY

Let us reconsider the synthesis of ammonia from hydrogen and nitrogen as described in Section 3.11, but performed in an open reactor. We may adjust the input and output of reactants and products into the reactor in such a manner that the direction of ammonia production is favored. Thus, to steady flows J_{H_2} and J_{N_2} of a given magnitude, there corresponds a well-defined value of J_{NH_3}. These flows may also be obtained by starting from a large quantity of ammonia and following the reaction as it progresses in the reverse direction. However, such behavior is not always the case with nonlinear chemical reactions operating far from the thermodynamic equilibrium state. For example, the BZ reaction does not always follow this simple route.

To see this, let us adjust the various fluxes of BZ reagents to a well-defined steady state value and measure an electrical property (redox potential) of the system which reflects the amount of substances present in the CSTR. We choose the concentration of $Na\,BrO_3$ as a constraint parameter and increase its value so that a new steady state is reached by the reaction. Figure 8.13 shows such successive measurements. At the beginning of the experiment the system follows branch (1) smoothly until point b is reached. A further increase of the value of the constraint beyond the critical value of $[Na\,BrO_3]_c$ provokes

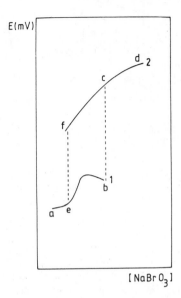

Fig. 8.13. In the BZ reaction the electrochemical potential as a function of $[NaBrO_3]$ shows a chemical hysteresis. (Adapted from P. De Kepper *et al.*, 1976)

smoothly on this new branch, traveling on the path *abcd*. Now we start from *d* and decrease $[NaBrO_3]$ in a backward journey toward point *a*. Experience shows that the representative point of the system travels along the path *dfea*, exhibiting the phenomenon of *hysteresis*. Thus the chemical system has an intrinsic memory; if it starts traveling from point *a*, it follows branch (1), whereas if it starts from point *d*, the second path is chosen by the system. Moreover, if the concentration of sodium bromate is held between *e* and *b*, two simultaneous stable pathways are open to the system. We speak of the appearance of *multiple steady states*. In this region, in a given experimental conditions, path (1) or path (2) may be chosen by the system indifferently.

8.8. A CHEMICAL REACTION MAY PERFORM MORE TRICKS

If the excitable and unstirred BZ reacting medium is arranged in a three-dimensional system, the spirals take the form of unfolding scrolls. Sometimes the scroll axis closes onto itself and amazing temporal structures arise. Figure 8.14 shows such a pattern.

An initially homogeneous BZ mixture may become unstable and form

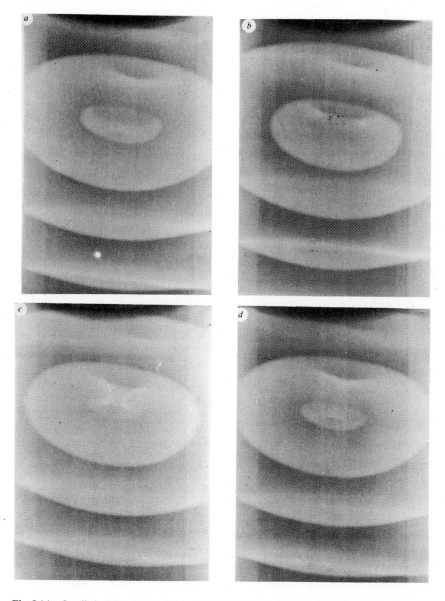

Fig. 8.14. Scrolls in BZ reaction. (From B. J. Welsh *et al.*, 1983, reprinted by permission from *Nature*, Vol. 304, p. 611. Copyright © 1983 Macmillan Journals Limited.)

mosaïc-type structures. These are stationary structures and do not propagate. At the present time their nature and origins are still not understood although convective phenomena clearly play a role.

*   *

*

The BZ reaction and its amazing properties have changed markedly our view of chemical dynamics. A chemical process is no longer only the mixing of some inert products to produce other "lifeless" components. Under far from equilibrium conditions, a necessary condition for living species, a chemical reaction becomes "alive." It measures time, it propagates information, it discriminates between past and future, between left and right. It may also construct mosaïc structures. If pushed too hard it starts to vacillate and behave erratically or "irrationally."

The BZ reaction is nothing more than four ordinary and simple chemical products dissolved in water. The extraordinary properties of the BZ reaction result either from chemical processes alone, or from the correlation between convective motion of water and chemical reactions. The presence of water molecules and, therefore, convective effects may be eliminated from the reacting medium in "model" or "on paper" chemical reactions. As we shall see, investigation of these model reactions by mathematical methods shows that all exotic properties of the BZ reaction may be recovered if we only consider chemical reactions and diffusion processes.

The complex processes of life are made up of superposition of many thousands of interrelated chemical reactions. Some of these surely operate far from thermodynamic equilibrium. Thus if one simple reaction may exhibit such diverse "lifelike" behavior, then we may expect that a large collection of such reactions alone, or combined with hydrodynamic phenomena, may create and sustain life.

*   *

*

In the next chapter we present a few simple chemical models that exhibit the strange behavior of the BZ reaction. We show how to analyze these model reactions using differential equations and how to detect the presence of self-organizational behavior in them.

Chapter 9

A MATHEMATICAL APPROACH TO THE NEW CHEMISTRY

9.1. INTRODUCTION

In Section 2.10 we saw that, given a reaction mechanism, a set of differential equations can be constructed which describe the time change of concentrations of reacting species. The integration of these equations yields functions which account for the time evolution of the chemical reaction.

One look at Fig. 8.4 tells us how difficult such an analysis can be. The great number of intermediate species of the BZ reaction necessitates an equally large number of coupled differential equations. For each equation, rate constants must be measured experimentally. Moreover, the mechanism of Fig. 8.4 is only partial; the details of some of its parts have not yet been elucidated.

Faced with such practical difficulties a theoretician must simplify his problem. He tries to determine the most important intermediate steps of the reaction mechanism, those which are likely to determine the behavior of the system. With these elements, he constructs an abstract mechanism, a model, much simpler than reality, but exhibiting the global properties of the original scheme. A model must be simple enough to be handled analytically or at least not require expensive computer analysis.

Model making is the essence of all theoretical approaches, and good models rely heavily on experimental knowledge which has been rearranged and condensed into a simpler form. Sections 9.2 and 9.3 are devoted to models which contain the elements or the salient features of the BZ reaction.

9.2. A MODEL FOR THE BZ REACTION: THE OREGONATOR

Field and Noyes, working in Oregon, conceived of a chemical reaction model which exhibits the relevant features of the BZ reaction. It was named after its place of origin to carry on the tradition of the earlier-named "Brusselator" which we shall discuss in the next section. In particular, the model accounts for the properties of chemical clocks. It was constructed by a careful analysis of the known experimental facts of the BZ scheme.

Experience shows that the concentrations of the initial and final products of the BZ reaction do not oscillate, but that only the intermediate products undergo periodic changes. Furthermore, as we shall see, a periodic behavior independent of the initial conditions cannot arise from linear differential equations. Thus, the differential equations of the model must exhibit nonlinear contributions with respect to the pertinent variables of the problem. Moreover, autocatalytic reactions introduce a "snowball" effect in the reaction medium which brings about the oscillatory behavior. Thus we may conclude that $HBrO_2$ must be one of the important variables of the model, since it is the switch element which pushes the reaction into either the first or the second box of Fig. 8.4. Ce^{4+} is another obvious variable that must be considered in any model of the BZ reaction.

In addition, the level of Br^- ions in the reaction mixture controls the change from one alternate path to the next, and thus the bromide ion concentration must also be considered as a variable. The Oregonator model is constructed precisely with these three variables and is represented in the following scheme:

$$A + Y \to X + P$$
$$X + Y \to 2P$$
$$A + X \to 2X + 2Z: \quad \text{autocatalytic step}$$
$$2X \to A + P$$
$$B + Z \to hY \tag{9.1}$$

The symbol A stands for BrO_3^-, B is bromomalonic acid, P is HOBr, X denotes $HBrO_2$, Z is Ce^{4+} and Y represents Br^- ions. X, Y, and Z are the time- and space-dependent concentrations, while A, B, and P are assumed to remain constant at all times, since they enter and leave the reaction medium at constant rates. Following the methods of Section 2.10, we see that three coupled differential equations are necessary for the description of the model, Eq. (9.1). If diffusion is of importance in the description of experimental observations (such as spiral and concentric wave propagation), then a contribution accounting for the diffusion processes must be added to the equations.

In recent years the Oregonator has been the focus of intensive theoretical investigations; many of the peculiar properties of the BZ reaction have been described by this simple model. However, from a mathematical point of view the Oregonator is still an extremely complex system, even if diffusion is not considered. The three relevant variables are related via a set of three nonlinear differential equations. Twentieth century mathematics is still unable to solve such problems in a general way, and various approximate methods must be used.

A detailed analysis of the Oregonator model is beyond the scope of the present book. However, quite similar properties may be found by analyzing another, still simpler, two-variable model which is easier to handle. This scheme is known as the *Brusselator*, and is certainly one of the most famous chemical reaction models.

9.3. THE BRUSSELATOR

9.3.1. The Well-Stirred Reactor

It is obvious that we need at least two variables, $X = X(t)$ and $Y = Y(t)$, to produce oscillations in a chemical system. If, for example, the concentration of X has the tendency to increase, it will continue to do so unless X reacts with Y. The latter might eventually decrease the concentration of X while increasing its own output. On the other hand, if too much Y favors the production of X, then the reaction is trapped in an oscillatory change of X and Y concentrations.

One can easily show that it is not possible to have oscillatory behavior in two-variable chemical systems if the elementary steps are unimolecular or bimolecular; the presence of a trimolecular step is necessary. Remembering also that autocatalysis is an obsolute requirement for the onset of oscillations, the following basic model may be constructed:

$$A \underset{k_{-1}}{\overset{k_1}{\rightleftharpoons}} X$$

$$B + X \underset{k_{-2}}{\overset{k_2}{\rightleftharpoons}} Y + D$$

$$2X + Y \underset{k_{-3}}{\overset{k_3}{\rightleftharpoons}} 3X$$

$$X \underset{k_{-4}}{\overset{k_4}{\rightleftharpoons}} E \tag{9.2}$$

The name *Brusselator* was given to the scheme of Eq. (9.2) by Tyson as a tribute to the fact that this model was developed in Brussels by Prigogine and Lefever in 1968. Since then, the scheme has been the focus of extensive analysis. The popularity of this minimal model stems from the fact that the time change of the concentrations of the intermediates X and Y exhibit the most relevant properties of the actual BZ reaction and also predicts other self-organizing behavior that, as yet, has not been conclusively demonstrated in any real chemical reaction.

The Brusselator is assumed to proceed in an open reactor (see Fig. 9.1). By appropriate adjustment of input fluxes J_A, J_B and output fluxes J_D, J_E, the concentrations of A, B, D, and E are held constant and far from their equilibrium values. These adjustable concentrations are the *constraints* of the system, and in the description of the model enter the differential equations as constant parameters. Intermediates X and Y are produced as a *response* to the constraints acting on the system. Their concentrations are regulated by the chemical process itself. Thus, the concentrations of X and Y may vary in time or in space, and therefore they form the variable quantities which enter the differential equations describing the kinetics of the scheme.

To each set of constraints, there corresponds a set of responses, and together they determine the state of the system. It is assumed that the reaction proceeds in a single liquid or gaseous phase, and that initially all concentrations are uniform throughout the reactor. These type of mixtures are referred to as *uniform systems*.

We now suppress all the input and output fluxes of the reactor and wait for the onset of chemical equilibrium. The law of mass action, Eq. (2.14), must be applied to each elementary step of the scheme, Eq. (9.2), separately as if they were independent chemical reactions. One finds immediately that

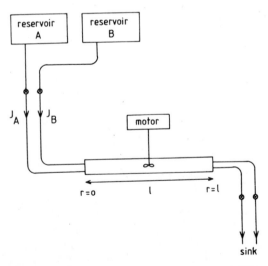

Fig. 9.1. This model reactor is a one-dimensional narrow tube of length l. Concentrations of A and B, are held constant at all times.

$$X^{\text{eq}} = \frac{k_1 A}{k_{-1}} \qquad Y^{\text{eq}} = \frac{k_{-3} k_1 A}{k_3 \, k_{-1}}$$

$$\frac{k_1 A}{k_{-1}} = \frac{k_{-4} E}{k_4} \qquad \frac{k_2 B}{k_{-2}} = \frac{k_{-3} D}{k_3} \tag{9.3}$$

Thus, at chemical equilibrium there exists a well-defined and unique relationship, independent of time and space, between the kinetic constants and the concentration of the initial and final products. We are thus confirming that a chemical system at equilibrium is "dead" and cannot show, for example, any concentration oscillations or wave propagation.

The mixture in the imaginary reactor may be kept at any required distance from equilibrium, provided fluxes of products or reactants are introduced into the system. In a more radical procedure, we may keep the model reaction infinitely far from equilibrium by suppressing the reverse reactions altogether. This method has the merit of considerably simplifying the rate equations. These equations may be simplified further, without prejudice to the properties of the model, if we choose a unit value for all rate constants. With these simplifications and with the help of the methods of Section 2.10 the differential equations for the Brusselator are:

$$\frac{dX}{dt} = A - (B + 1)X + X^2 Y$$

$$\frac{dY}{dt} = BX - X^2 Y \tag{9.4}$$

The set of Eqs. (9.4), together with initial values of X and Y, specify completely the future evolution of concentrations inside the reactor.

By a suitable adjustment of the constraints, the system can be held at a *steady nonequilibrium state*, in which case $dX/dt = 0$ and $dY/dt = 0$. Whenever such conditions are imposed to the differential equations (9.4) the solution X^s, Y^s is called *a singular point*.

A straightforward stability analysis, based on the thermodynamic methods of Section 6.4, shows that the excess entropy production of scheme (9.2) around a given nonequilibrium steady state $X = A$ and $Y = B/A$ is

$$\delta_X P = \sum_{i=1,4} \delta v_i \delta A_i = R \left[(1 - B)\frac{(\delta X)^2}{A} + \frac{A^s}{B}(\delta Y)^2 \right] \tag{9.5}$$

Thus, for sufficiently large values of product concentration $B = B_c$, $\delta_X P$ may become negative. This fact confirms the instability of the steady homogeneous state $X^s = A$, $Y^s = B/A$. We can conclude that if the Brusselator scheme is held far from thermodynamic equilibrium, nonuniform and time-dependent behavior becomes possible. The exact nature of this time dependence is beyond the scope of nonequilibrium thermodynamics. We may expect time-periodic solutions indicating the possible presence of oscillating chemical reactions.

9.3.2. The Unstirred Brusselator

In Section 8.3 the qualitative description of band propagation in the BZ reaction was based upon the diffusion of $HBrO_2$ molecules, and therefore the diffusion process must somehow enter into a dynamic description of the reaction mixture.

In the description of chemical waves, we assume that the concentrations change in time not only through the reaction processes but also because of diffusion of substances into or out of the unit volume under consideration. The simplest way to account for the diffusion of X and Y is to assume that they follow Fick's law, with constant diffusion coefficient D as discussed in Section 4.9. Moreover, to avoid unnecessary and cumbersome computations, we assume that the imaginary reactor is a long and narrow tube (see Fig. 9.1) so that the nonhomogeneous distribution of concentrations is restricted to one dimension, along the tube, taken as the r axis. We therefore assume a one-dimensional diffusion process, and the differential equations read

$$\frac{\partial X}{\partial t} = A + X^2 Y - (B + 1)X + D_X \frac{\partial^2 X}{\partial r^2}$$

$$\frac{\partial Y}{\partial t} = BX - X^2 Y + D_Y \frac{\partial^2 Y}{\partial r^2}$$

$$0 < r < l \tag{9.6}$$

These two coupled *reaction-diffusion equations* are nonlinear partial different-ial equations of the second order and of the *parabolic* type, whereas Eq. (9.4) form two coupled ordinary, nonlinear, first-order differential equations.

The desired solutions $X(r, t)$ and $Y(r, t)$ of Eq. (9.6) can only be determined if their behavior at the two ends is prescribed. In other words, we must specify the *boundary conditions* of the problem. Two classes of boundary conditions are most appropriate for the study of the chemical reactions.

1. There is no leakage of X and Y into the external medium and the system operates under *zero-flux boundary conditions*. As the flux of X or Y at the boundaries of the tube is described by the gradient of the latter along the r

coordinate, zero-flux boundary conditions imply that

$$\left(\frac{dX}{dr}\right)_{r=0} = \left(\frac{dY}{dr}\right)_{r=0} = \left(\frac{dX}{dr}\right)_{r=l} = \left(\frac{dY}{dr}\right)_{r=l} = 0 \qquad (9.7)$$

In some problems, we may keep artificially $X^s = A$ and $Y^s = B/A$ at the boundaries of the system.

If the reaction-diffusion equations are solved, subject to one of the boundary conditions cited above, the solutions $X = X(r, t)$ and $Y = Y(r, t)$ give the local values of the concentrations of these two intermediates in time. Our macroscopic knowledge of the chemical system is complete.

With the example of the BZ reaction, we saw that all self-organizing phenomena are crucially dependent on the values of the constant quantities of the various reagents in the reactor. This fact is reflected in the solution of Eqs. (9.6) by the appearance of A, B and also the various rate constants as parameters in the solutions $X = X(r, t, A, B)$ and $Y = Y(r, t, A, B)$.

As in the BZ reaction, nonequilibrium conditions alone are not sufficient for the appearance of spatial and temporal organization in the Brusselator model. Time and space organization appears only for well-defined parameters A, B, D_X, and D_Y.

Eqs. (9.4) and (9.6) look deceptively simple. As yet, no general method for the solution of a set of nonlinear ordinary or partial differential equations exists. Only a few exceptional cases have been solved analytically. Faced with the near-impossibility of a direct and exact solution, one is forced into numerical integration, or approximate solutions. Qualitative methods also have been developed which predict the approximate nature of solutions. The object of the next three sections is to give a short summary of these methods.

By now, the reader must be bewildered by our slashing simplifications. He may ask himself, and rightly, how we can hope to describe the complex scheme of Fig. 8.4, already only one part of the physical reality, by a simple, two-variable system chosen partly for convenience? The answer is that Eqs. (9.4) and (9.6) and those we could have obtained by considering all the intermediates of Fig. 8.4 belong to the same general class of mathematical objects. Therefore, their respective solutions are qualitatively similar, and at this stage we are only interested in qualitative properties. The similarity of solutions of a given class of differential equations is an important element of model construction.

Were we able to solve Eqs. (9.6) directly, we would have obtained expressions of the type $X = X(t, r, A, B)$ and $Y = Y(t, r, A, B)$. For example, starting from the equilibrium values, by varying the parameter B we could verify directly that at $B = B_c$, the homogeneous solution is unstable. Beyond this critical value, new solutions would emerge, exhibiting time or space-time periodicities. Unfortunately, this direct line of approach is currently not possible, and one is forced into devising indirect and approximate methods which are the basis of what is known as *bifurcation theory*.

9.4. BIFURCATION THEORY

From the thermodynamic stability criterion of Chapter 6, we expect that the rate equations, (9.4) and (9.6), imply a unique, spatially uniform and temporally monotonic, solution in the vicinity of the equilibrium state. However, at some distance from the equilibrium state, we expect that this thermodynamic solution becomes unstable, and suddenly new solutions appear which manifest temporal or spatio-temporal periodicities.

If such solutions are possible, then we may have accounted in mathematical language for the strange behavior of the BZ reaction. The mathematical problems raised by the solution of nonlinear differential equations exhibiting the properties cited above, lead to a branch of mathematics called *bifurcation theory*. This term stems from the fact that the unique solution existing in the immediate neighborhood of the equilibrium domain reaches a point, called the *bifurcation point*, where one or several new possibilities are open to the system. This situation is analogous to the choice of a jogger who leaves his home and reaches an intersection of three possible roads (see Fig. 9.2). The road

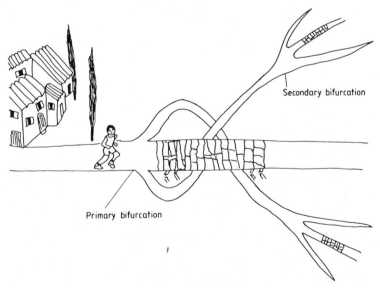

Fig. 9.2. Visualization of primary and secondary bifurcations.

continues over a shaky bridge. If he follows the bridge, he loses his balance and falls into either of the two "stable" roads.

Bifurcation theory was initiated and developed by Poincaré, Andronov, Hopf, and many others. Currently the theory is the focus of much attention and is one of the most active fields of applied mathematics. It has innumerable applications in many diverse fields of science, ranging from physics and chemistry to economics and the social sciences.

The aim of bifurcation theory is twofold.

1. For a given problem, one demonstrates analytically the existence of branching points for certain values of parameters. In our example, the road length is the bifurcation parameter.
2. Near the bifurcation point, one constructs approximate solutions for the newly emerging reaction pathways.

In what follows, we shall concentrate mainly on the first point. The construction of new branching solutions is a lengthy and very involved procedure, and is beyond the scope of this book.

The theory of bifurcation is intimately related to the important concept of the stability of the thermodynamic branch (solutions to kinetic equations far from equilibrium which are the continuation of equilibrium values). It can be shown that for any given parameter, if after a first bifurcation a stable solution becomes unstable and branches out into several solutions, according to the principle of the exchange of stability at least one new stable state must appear. This principle states the obvious fact that since the system exists, it must be in some state.

Determination of the parameters at which the branching occurs is of primordial importance for both the analytical and numerical solutions of nonlinear differential Eqs. (9.4) and (9.6). Usually the approximate analytical solutions are constructed only in the immediate neighborhood of the branching points. Therefore, the position of the latter must be determined. Moreover, we have already seen that, for a given reaction mechanism, self-organizational properties only appear in a well-defined and narrow range of parameters. If this range is not determined by an independent method, numerical integration of the differential equations becomes extremely difficult or even impossible. Therefore, the first task in any search for bifurcating solutions is the determination of the unstable points of the homogeneous system. Thus, we must go back to the stability analysis of the various states. However, it is more appropriate to test the stability of differential equations by kinetic methods. We shall devote the next section to a brief outline of the methods of *linear stability* or *normal mode* analysis.

The concepts and methods of linear stability analysis are of a "technical" nature. The uninterested reader may proceed directly to Section 8. He should only remember that unstable steady states may be detected easily.

9.5. THE CONCEPT OF STABILITY REVISITED

9.5.1. Linear Stability Analysis in a Well-Stirred Reaction

In Chapter 6 we defined the concept of stability in a macroscopic system. We also developed a thermodynamic stability criterion for testing the stability properties of systems that are far from equilibrium. But, before we go any further, let us define the concept of *perturbation* that will be used a great deal in what follows. In a well-stirred vessel, the perturbation imposed on a system may be defined as the addition of a small amount of one of the intermediates (for example Br$^-$ ions in the BZ reaction) into the reaction vessel. The addition of the chemical is assumed to be an instantaneous, single-step event which does not alter the constraints imposed on the system.

A perturbation may be endogenous. For example, at a given point in the system a momentary and spontaneous small increase or decrease in the local concentration of one or several of the intermediates may appear. These are internal perturbations, or *fluctuations*, of the system. Since near critical points both fluctuations and perturbations have the same destabilizing influence, we shall use both terms interchangeably.

The stability properties of steady states can be tested by the analysis of the dynamic differential rate equations, (9.4) and (9.6). If one studies the response of a given steady state to random fluctuations, one finds that the fluctuations take the system momentarily away from the steady state. Therefore, the state of the system is again time-dependent and has the tendency to evolve according to Eqs. (9.4) and (9.6). If, in the course of time, the new time-dependent state tends toward the original steady state, the latter is stable. For an unstable steady state, random fluctuations are amplified and drive the system away from the original state and into new modes of behavior.

The first step in any stability analysis is the determination of the undisturbed steady states. In chemical systems, the search for the steady states amounts to solving simultaneous algebraic equations which are obtained by setting the time change of concentrations in Eqs. (9.4) and (9.6) equal to zero. Often this means that one must solve a polynomial equations of the nth degree. For the Brusselator scheme we find

$$X^s = A \quad Y^s = \frac{B}{A} \tag{9.8}$$

Therefore, once the constraints are given, the steady states are determined.

If X^s and Y^s are the known steady states, and if $x(t)$ and $y(t)$ are random fluctuations, then the new states $X = X^s + x(t)$ and $Y = Y^s + y(t)$ must evolve according to the following set of differential equations:

$$\frac{dx}{dt} = -(B+1)x + Y^s(2X^sx + x^2) + y(X^{s^2} + 2xX^s + x^2)$$

$$\frac{dy}{dt} = Bx - Y^s(2X^sx + x^2) - y(X^{s^2} + 2xX^s + x^2) \qquad (9.9)$$

The solution of Eq. (9.9) provides the explicit time behavior of $x(t)$ and $y(t)$, and therefore determines the stability properties of the steady state. These equations are as difficult to solve as the original Eq. (9.4) from which they have been derived. However, if X^s and Y^s are only slightly disturbed—that is, we are only probing the immediate neighborhood of a steady state (see Fig. 9.3) then $(x(t)/X^s) \leqslant 1$ and $(y(t)/Y^s) \leqslant 1$. Thus, the two polynomials may be expanded in a Taylor's series around the values X^s and Y^s. For sufficiently small values of x and y, only first-order terms must be considered. In this approximation, the nonlinear rate equations, (9.9), reduce to a set of two linear, coupled differential equations,

$$\frac{dx}{dt} = -(B+1)x + 2Y^sX^sx + yX^{s^2}$$

$$\frac{dy}{dt} = Bx - 2xY^sX^s - yX^{s^2} \qquad (9.10)$$

Fortunately, these equations may be solved easily. We can verify immediately,

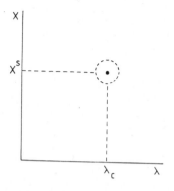

Fig. 9.3. The linear stability analysis around a solution X^s is only valid in a small region surrounding the singular point.

by direct insertion, that functions of the form $x = x_0 e^{\omega t}$ and $y = y_0 e^{\omega t}$, are particular solutions of Eq. (9.10), provided x_0, y_0, and ω are chosen appropriately. By inserting x and y into Eqs. (9.10), we obtain two homogeneous algebraic equations of the first degree in x_0 and y_0; ω appears as a parameter in the equations

$$\omega x_0 = (B - 1)x_0 + A^2 y_0$$
$$\omega y_0 = - Bx_0 - A^2 y_0 \qquad (9.11)$$

We know from elementary algebra that the condition for the existence of a nontrivial solution for x_0 and y_0 is that the determinant of the coefficients in Eq. (9.11) vanishes.

$$\begin{vmatrix} \omega - B + 1 & -A^2 \\ B & \omega + A^2 \end{vmatrix} = 0 \qquad (9.12)$$

The determinant in its expanded form furnishes a second-degree algebraic equation in the parameter ω,

$$\omega^2 + (A^2 - B + 1)\omega + A^2 = 0$$

or

$$\omega^2 - T\omega + \Delta = 0 \qquad (9.13)$$

This *characteristic equation* is solved readily and yields two values ω_1 and ω_2, given by

$$\omega_{1,2} = \frac{T \pm \sqrt{T^2 - 4\Delta}}{2} \qquad (9.14)$$

One verifies easily that for any second-degree equation, T is always the sum and Δ the product of the two roots ω_1 and ω_2.

Therefore, $x = x_0 e^{\omega_1 t}$ and $x = x_0 e^{\omega_2 t}$ are both solutions to Eq. (9.10). In the theory of differential equations one shows, and the reader may verify readily by direct insertion, that the most general solution of Eq. (9.10) is

$$x = C_1 e^{\omega_1 t} + C_2 e^{\omega_2 t}$$
$$y = C_1 K_1 e^{\omega_1 t} + C_2 K_2 e^{\omega_2 t} \qquad (9.15)$$

Usually in such equations the constants C_1 and C_2 are determined by imposing the values of the functions at time $t = 0$. We say that Eq. (9.10) must be solved, subject to initial conditions.

It can be shown easily that the constants K_1 and K_2 are found by solving another second-degree equation, namely,

$$A^2 K^2 + (B - 1 + A^2)K + B = 0$$

As we shall see in Section 9.6, the linear stability analysis of the steady states of a given chemical scheme may be achieved without the explicit knowledge of functions $x(t)$ and $y(t)$. All we need is the time behaviors of these functions, which are dictated by $e^{\omega_1 t}$ and $e^{\omega_2 t}$. Thus the linear stability analysis reduces to the investigation of the nature of the roots of the characteristic equation (9.13).

In the absence of a rigorous mathematical study of solutions of coupled nonlinear differential equations, linear stability analysis is a powerful tool for detecting the eventual presence of unstable solutions, and therefore the onset of self-organization in the system. Now we extend the methods of linear stability analysis to the case of reaction-diffusion equations.

9.5.2. Stability Analysis in the Presence of Diffusion

Following exactly the methods of the preceding section, with the help of the linear stability analysis the unstable points of the reaction-diffusion equations, (9.6), may be found. As the contribution of the diffusion is already linear, the linearized equations are

$$\frac{\partial x}{\partial t} = (B - 1)x + A^2 y + D_X \frac{\partial^2 x}{\partial r^2}$$

$$\frac{\partial y}{\partial t} = -Bx - A^2 y + D_Y \frac{\partial^2 y}{\partial r^2}$$

$$0 < r < l \tag{9.16}$$

However, now the small disturbances $x = x(t, r)$ and $y = y(t, r)$ are not only time- but also space-dependent. Because of the presence of the diffusion process, the influence of the fluctuations is crucially dependent on the location of their action on the system. Moreover, in the presence of diffusion, the perturbed solution $X(t, r) = X^s + x(t, r)$ is subject to the same boundary conditions as the original nonlinear equations. Thus, for fixed boundary conditions $x_0(r, t) = x_l(r, t) = 0$, whereas zero-flux boundary conditions require that

$$\left(\frac{\partial x}{\partial r}\right)_{r=0} = \left(\frac{\partial x}{\partial r}\right)_{r=l} = \left(\frac{\partial y}{\partial r}\right)_{r=0} = \left(\frac{\partial y}{\partial r}\right)_{r=l} = 0 \tag{9.17}$$

It can be proved easily, by direct insertion, that solutions of the type $e^{\omega t} \sin(m\pi r/l)$, $m = 1, 2 \ldots$ or $e^{\omega t} \cos(m\pi r/l)$, $m = 0, 1, 2 \ldots$ satisfy Eq. (9.16) for fixed and zero-flux boundary conditions, respectively.

The set of coupled differential equations (9.10) and (9.16) are members of the same class of mathematical objects, and thus are analyzed by the same procedure. Presently the characteristic equation is computed from the determinant

$$\begin{vmatrix} \omega - B + 1 + D_X \dfrac{m^2\pi^2}{l^2} & -A^2 \\[2mm] B & \omega + A^2 + D_Y \dfrac{m^2\pi^2}{l^2} \end{vmatrix} \qquad (9.18)$$

The supplementary contribution $D_X m^2\pi^2/l^2$ and $D_Y m^2\pi^2/l^2$ arise from the diffusion process, through $\partial^2(\cos(\pi m r/l))/\partial r^2 = -(m^2\pi^2/l^2)\cos(\pi m r/l)$. The characteristic equation is found by the same procedure as before. However, the quantities T and Δ are now functions of the diffusion coefficients and the length of the system.

9.6. PHASE PLANE ANALYSIS

In this section we develop a procedure for testing the stability of a system without the necessity of explicitly evaluating the quantities $x(t)$ and $y(t)$. We can illustrate the method with a simple example taken from mechanics.

Consider an old, wound-down clock. Its pendulum is hanging vertically in an equilibrium position, as shown in Fig. 9.4a. If a momentum is given to the pendulum, it swings for a while but finally comes to a stop again at its previous equilibrium position. Two forces dampen the motion of the pendulum. These are the frictional forces at the intersection of the pendulum and the main body of the clock, and the viscous forces generated by the motion of the pendulum in the air. A part of the mechanism inside the clock is designed to compensate for these frictional forces. In an ideal frictionless world, once the pendulum starts swinging, it continues to do so regularly, without ever stopping.

Another function of the complex clock mechanism is to insure that the pendulum always swings 60 times per minute. This property is independent of the magnitude of the initial force which puts the clock in motion.

There is a fundamental difference between the dynamics of a clock and an ideal pendulum. The same qualitative difference exists also between the oscillating BZ reaction and the periodic changes in an ideal prey-predator population. The internal mechanism of the BZ reaction operates like a clock, whereas prey-predator systems exhibit dynamics that show pendulum-like

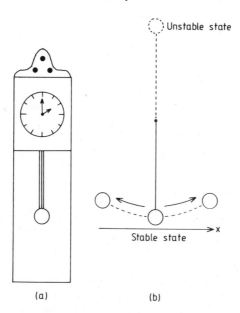

Fig. 9.4. (a) By analogy, the overall motion of the clock may be represented by a limit cycle. (b) Stable and unstable positions of an ideal pendulum.

properties. We must note, however, that the similarities mentioned above are purely formal.

9.6.1. Motion of a Pendulum

Figure 9.4b shows an ideal pendulum of mass m. Let F be the force that tends to bring the pendulum from point x back to its resting position $x = 0$. In addition, we assume that this force only generates small angle oscillations. The force is therefore given by $F = -kx$. The constant proportionality coefficient k and the mass m specify the individuality of the pendulum. The state of the pendulum is completely determined by two independent variables, namely the position coordinate $x = x(t)$ and the velocity function $v = v(t)$. These two functions are related in the following manner:

$$\frac{dx(t)}{dt} = v(t), \qquad \frac{dv(t)}{dt} = -\frac{kx}{m} \tag{9.19}$$

The first equation defines the velocity function, and the second expresses Newton's law as applied to the oscillating pendulum. One verifies immediately

that

$$x(t) = K \cos(\omega t + \alpha), v(t) = -K\omega \sin(\omega t + \alpha) \qquad (9.20)$$

are solutions to the set of Eq. (9.19) and completely account for the motion of the pendulum. The *frequency*, $\omega/2\pi = (1/2\pi)\sqrt{k/m}$, is a measure of the number of oscillations per unit of time and is determined by the acting force and the mass of the pendulum. The constants K and α are calculated from the knowledge of the given initial values of position and velocity of the pendulum. For example if at $t = 0$, $x = 2$, and $v = 1$, then K and α are computed from the two coupled equations $2 = K \cos \alpha$ and $1 = -K\omega \sin \alpha$.

Figure 9.5 shows the time evolution of the state of the pendulum as given by Eq. (9.19). As the pendulum moves back and forth, the *representative point* of the system moves along an upward winding spiral. The projection of the system's trajectory on the $x - t$ and $v - t$ planes shows two sinusoidal functions as given by Eq. (9.20).

In some instances, we might only want to know how the velocity of the pendulum changes at different points along the trajectory. That is to say, we would like to construct a function $v = v(x)$. This may be achieved by eliminating time from Eq. (9.20). As $\cos^2 \alpha + \sin^2 \alpha = 1$, we find that

$$\frac{x^2}{K^2} + \frac{v^2}{\omega^2 K^2} = 1 \qquad (9.21)$$

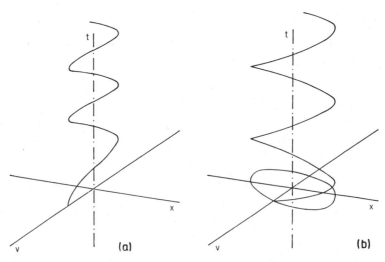

Fig. 9.5. (*a*) The change in time of v. The trajectory of the pendulum in the three-dimensional space of $x, v,$ and t.

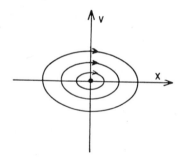

Fig. 9.6.   The phase plane trajectories of the ideal pendulum for three different initial conditions.

For given initial conditions, or equivalently, for a fixed value of K, Eq. (9.21) represents an ellipse in the $v - x$ plane as seen in Fig. 9.6. To each value of K there is a corresponding ellipse.

The $v - x$ coordinates define a two-dimensional *phase plane* and an ellipse of the type of Eq. (9.21) is the *phase trajectory* of the pendulum for the specified conditions. Phase trajectories are the projection of the three-dimensional spiral on the $v - x$ plane. In other words, it is the curve we see if we look at the spiral from the positive t direction. Phase trajectories are the two-dimensional portrait of the three-dimensional spiral, and the information contained in the third dimension is absorbed in the direction of rotation of the phase trajectories.

As we shall see, the concept of phase trajectories is very important for the analysis of coupled nonlinear differential equations. In the absence of a complete analysis of a dynamical system, the phase trajectories give us some indication about the nature of the motion.

9.6.2. The Reverse Problem

In the preceding example the phase trajectory of the motion is found by solving the dynamical equations of motion of the pendulum, and the truncated information that it provides is not of much interst. However, if the phase plane trajectories of dynamical systems could be constructed by some independent means, information about the three-dimensional motion might be obtained. Thus, the determination of phase trajectories may be a method for the investigation of the stability of nonlinear differential equations.

We shall illustrate the methods and the terminology of phase space analysis by considering again the simple problem of the pendulum, pretending that Eq. (9.19) cannot be solved directly. Time can be eliminated from these equations at once by dividing one of them by the other. We get:

$$\frac{dv}{dx} = -\omega^2 \frac{x}{v} \qquad (9.22)$$

If this equation can be integrated analytically, we get $v = v(x)$, which is called the *integral curve* and is precisely the phase portrait of the motion of the pendulum. The reader may check, by direct integration, that the integral curves are ellipses, as expected. Thus, by going backward in our thinking, we guess that the representative point of the pendulum moves along the spiral of Fig. 9.5.

9.6.3. Properties of the Phase Plane

When the representative point of the pendulum travels in the three-dimensional space of x, v, and t on a spiral trajectory, its image on the phase plane moves along one of the ellipses of Fig. 9.6. The velocity and direction of this motion give an indication of the "lost" variable t. It is obvious that the velocity of the representative point in the phase plane has two components: dx/dt in the x direction, and dv/dt in the v direction. Both quantities are given by Eq. (9.19). In the phase plane of Fig. 9.6 the motion occurs clockwise, as $dx/dt = v$ is positive in the upper half plane and x increases with time.

From these consideration we see that the shape of the phase trajectories provides an indication of the nature of the qualitative solutions of the original differential equations.

9.6.4. Stability Analysis

We go back to the ideal, frictionless pendulum hanging in its vertical position and characterized by $x = 0$ and $v = 0$. With a small momentary displacement $x = \delta x$ the pendulum starts swinging and the motion is described by Eq. (9.19). If the solution $x = x(t)$ and $v = v(t)$ decreases in time and finally vanishes, then the position $x = 0$, $v = 0$ of the pendulum is a stable state. We say it is a stable *singular point*. Therefore, Eq. (9.19) serve the same purpose for the determination of the stability character of the singular points of a pendulum, as Eq. (9.10) in determining the stability of the steady states of the Brusselator.

Following the methods of Section 9.5, the characteristic equation of the pendulum is

$$\omega^2 = -1$$
$$\omega_{1,2} = \pm\sqrt{-1} \qquad (9.23)$$

Thus, the complete time-dependent solution $x(t)$ and $v(t)$ may be found by inserting these values of ω into Eq. (9.15). However, a full analytical form of the solution is not needed. From the time-dependent factors $\exp(\pm i\omega t) = \cos \omega t \pm i \sin \omega t$, we guess that the coordinates x and v of the representative point of the pendulum in the phase plane change periodically in time. Therefore, the phase trajectories necessarily are closed curves surrounding the singular point. Once the system leaves the $x = 0$, $v = 0$ point, it never returns to the original state again.

For every initial force exerted on the pendulum, a new closed phase trajectory (see Fig. 9.6) describing a new motion appears, and the ideal pendulum oscillates with regularity if a new perturbation does not interfere with its motion. We say that the singular point is *marginally stable*.

From this simple example, we can see that the knowledge of the shape of the phase trajectories is sufficient for the determination of the stability properties of a given system. We shall now use this method to investigate the Brusselator scheme.

9.7. PHASE PLANE ANALYSIS OF THE BRUSSELATOR

The nonlinear Eq. (9.4) is of the general type

$$\frac{dX}{dt} = f(X, Y)$$

$$\frac{dY}{dt} = g(X, Y)$$

If f and g are linearized around a given steady state X^s, Y^s one finds

$$\frac{dx}{dt} = ax + by$$

$$\frac{dy}{dt} = cx + dy \qquad (9.24)$$

where

$$a = \left(\frac{\partial f}{\partial X}\right)_s, \qquad b = \left(\frac{\partial f}{\partial Y}\right)_s, \qquad c = \left(\frac{\partial g}{\partial X}\right)_s, \qquad \text{and} \quad d = \left(\frac{\partial g}{\partial Y}\right)_s$$

Thus the analysis that follows is quite general and may be applied to any two-variable system.

If we eliminate time from Eq. (9.24), the resulting differential equation describes the phase trajectories in the x–y phase plane.

$$\frac{dy}{dx} = \frac{cx + dy}{ax + by} \tag{9.25}$$

The solution of this equation provides the desired function $y = y(x)$.

However, the phase trajectories may be found much more easily by considering Eq. (9.15), which is the most general solution to any two coupled linear differential equations. Such equations may be considered as two algebraic expressions defining the two unknown variables $e^{\omega_1 t}$ and $e^{\omega_2 t}$. Thus we find at once

$$e^{\omega_1 t} = \frac{y - K_2 x}{c_1(K_1 - K_2)}$$

$$e^{\omega_2 t} = \frac{y - K_1 x}{c_2(K_2 - K_1)}$$

from which we get

$$\left(\frac{y - K_2 x}{c_1(K_1 - K_2)}\right)^{|\omega_2/\omega_1|} = \frac{y - K_1 x}{c_2(K_2 - K_1)} \tag{9.26}$$

Equation (9.26) is a relation between y and x and is of the form $y = y(x)$; it describes the trajectories in the phase plane. The phase plane trajectories as deduced from this equation may be classified into various distinct groups according to the nature and the sign of ω_1 and ω_2. These phase trajectories may be deduced either from analytical considerations or by direct numerical computation.

9.7.1 Both Roots Are Real

From Eq. (9.14) one sees immediately that the two roots of the characteristic equation become real only if $T^2 - 4\Delta \geq 0$. In addition, if $\Delta > 0$ then $(T^2 - 4\Delta)^{1/2}$ is always smaller than T, and both roots have the same sign, which is determined by the sign of T. If T is positive, from Eq. (9.14) we deduce that both roots are positive. Therefore, all normal modes $e^{\omega t}$ are increasing functions of time, and thus the perturbations increase in time and the system never goes back to its original steady state. This fact is confirmed by phase plane analysis. It can be shown analytically that the phase trajectories of such systems are curves that all originate from the steady states $x = 0$, $y = 0$, and extend toward larger values of x and y. Thus the representative point of the

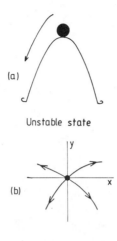

Unstable state

Unstable node

Fig. 9.7. The singular point is an asymptotically unstable node. All trajectories of the phase plane diverge from it.

system moves away from the singular point, indicating the instability of the steady state. The phase trajectories are shown in Fig. 9.7b. Such singular points are called *unstable nodes*.

If T is negative, from Eq. (9.14) we see that both roots are negative. The phase trajectories are the same as before. However, for all trajectories the

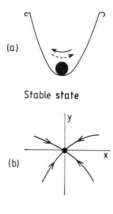

Stable state

Stable node

Fig. 9.8. The singular point is an asymptotically stable node. All trajectories of the phase plane are attracted toward it.

representative point of the system travels toward the singular point, as shown in Fig. 9.8b. This property stems from the fact that $e^{-\omega_1 t}$ and $e^{-\omega_2 t}$ both decay for large values of t. Thus, the steady state is stable and we speak of a *stable node*.

The stability concept of singular points may be visualized by analogy with examples taken from common experience. However, one must always keep in mind that this procedure is purely an aid to understanding, and does not imply identical physical processes.

Let us take a punch bowl and a small spherical ball. No matter where on the inside of the bowl we put the ball, it will invariably fall to the bottom (see Fig. 9.8a. In chemical reactions subject to a stable steady state regime, it all happens as if the system had constructed an imaginary punch bowl around the singular point (the steady state), which attracts all representative points that may appear inside the bowl. Now if we turn the punch bowl upside down, as shown in Fig. 9.7a, whatever the position of the ball it will roll down and disappear from the surface of the container.

If $\Delta < 0$ then $(T^2 - 4\Delta)^{1/2}$ is greater than T, and thus from Eq. (9.14) we conclude that the two roots ω_1 and ω_2 are necessarily of the opposite sign. The phase plane trajectories in this case are shown in Fig. 9.9a, and are hyperboloid curves with two asymptotes passing through the singular point. The asymptotes are called the *separatrixes* of the phase plane. We see that only if the representative point travels along the axis OA, will it go back to the steady state. All the other trajectories go to infinity as $t \to \infty$. This behavior stems from the fact that one of the normal modes is of the form $e^{\omega t}$ and increases indefinitely in time, driving the system away from the singular point.

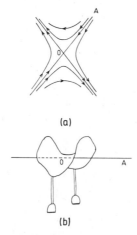

(a)

(b)

Fig. 9.9. Separatrixes and trajectories around a saddle point. The representative point on all trajectories but for one of the asymptotes go to infinity.

The steady state of the system is thus unstable. In this case, chemical reactions force the representative point of the system to move over a saddle-like surface, as shown in Fig. 9.9b. Therefore the singular point is called a *saddle point*.

An expert hand may roll a ball along the long axis of the saddle (OA) and make it stop at the singular point O. But the slightest knock causes the ball to move away from this state. On the other hand, it is impossible to reach the singular point from any other part of the saddle. The ball always falls and disappears. Thus singular points of the saddle type are always unstable.

9.7.2. The Roots Are Equal

If $T^2 - 4\Delta = 0$, from Eq. (9.14), we see that the two roots are identical and are given by $\omega_{1,2} = T/2$. For the particular case $a = d$ and $b = c = 0$, the phase trajectories are half lines which converge toward the singular point if $T < 0$. On the contrary, the trajectories leave the steady state if $T > 0$. We speak, respectively, of stable and unstable *stellar nodes*. These trajectories are shown in Fig. 9.10. In the general case, we are in the presence of a stable or unstable one-tangent node.

9.7.3. The Roots Are Complex Conjugates

If $T^2 - 4\Delta < 0$, we find that $\omega_1 = \omega_r + i\omega_i$ and $\omega_2 = \omega_r - i\omega_i$. Thus

$$e^{\omega_r t} e^{\pm i\omega_i t} = e^{\omega_r t}(\cos \omega_i t \pm i \sin \omega_i t)$$

and are functions with a time periodicity which is modulated by an exponential function. The behavior of the latter determines the final destiny of the system. If we note that $\omega_r = T/2$, then three cases are possible.

1. If $T > 0$, the phase plane trajectories are spirals that uncoil from the singular point. The steady state is said to be an *unstable focus* and is

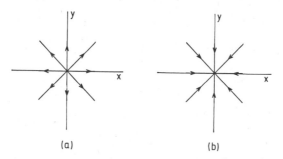

(a)   (b)

Fig. 9.10. (*a*) Asymptotically unstable stellar node. (*b*) Asymptotically stable stellar node.

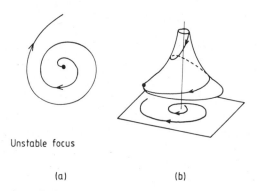

Unstable focus

(a)   (b)

Fig. 9.11. Asymptotically unstable focus. Spiral trajectories leave the singular point.

shown in Fig. 9.11a. In this case, $e^{\omega_r t}$ increases exponentially with time and the total solution departs from the singular point with an oscillatory motion.

2. If $T < 0$, the phase plane trajectories are again spirals, but they approach the singular point, as shown in Fig. 9.12a. Thus the latter is said to be a *stable focus*. In this case $e^{-\omega_r t}$ decreases with time and the total solution approaches the steady state in an oscillatory motion.

In both cases, it all happens as if the trajectories of the representative points of the systems, are constrained to move on the surface of an imaginary horn. Moreover, each chemical reaction determines the particular shape and size of the instrument.

Let us visualize the various events with the help of a real horn and a small metallic ball. The narrow end of the instrument represents the singular point and the small ball stands for the representative point of the system. We hold

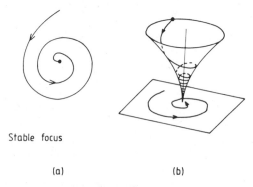

Stable focus

(a)   (b)

Fig. 9.12. Asymptotically stable focus. Spiral trajectories converge toward the singular point.

T	Δ	$T^2 - 4\Delta$	ω_r^1	ω_i^1	ω_r^2	ω_i^2		Phase Trajectories	Stability
+	+	+	+	0	+	0	Real		Unstable node
−	+	+	−	0	−	0	Real		Stable node
−	+	0	−	0	−	0	Real		Stable stellar node
+	+	0	+	0	+	0	Real		Unstable stellar node
−	−	+	+	0	−	0	Real $\}$		Unstable saddle point
+	−	+	−	0	+	0	Real $\}$		
−	+	−	−	$+i\omega$	−	$-i\omega$	Imaginary		Stable focus
+	+	−	+	$+i\omega$	+	$-i\omega$	Imaginary		Unstable focus
0	+	−	0	$+i\omega$	0	$-i\omega$	Imaginary		Center marginal stability

Fig. 9.13. Linear stability analysis. Singular points of a two-variable system and their stability.

the instrument in an upright position and put the ball inside the horn, as shown in Fig. 9.12b. No matter the initial position and velocity of the ball, it always rolls in a downward spiral trajectory and comes to rest at the bottom of the instrument. Thus the singular point of the system is a stable focus. We see also that the phase portrait in the $x-y$ plane is an inward coiling spiral.

In order to mimic the trajectories which originate from an unstable focus, we must turn the horn upside down and roll the ball over its external surface, as shown in Fig. 9.11b. Again, the ball rolls downward in an unwinding spiral and finally disappears. In this position, the singular point represented by the narrow end of the horn is an unstable focus.

3. If $T = 0$ then the two roots are purely imaginary: $\omega = \pm i\omega_i$, and we have $\cos \omega t \pm i \sin \omega t$. Thus the representative point in phase space undergoes undamped oscillations. Trajectories in phase space are closed curves surrounding a singular point which is referred to as a *center*. The ideal pendulum is an example of such a case (see Fig. 9.6).

In Fig. 9.13 we have compiled all the possible singular points of a set of two coupled differential equations, together with their stability properties.

9.7.4. Bifurcation Points of the Brusselator

In the Absence of Diffusion. Let us first consider the Brusselator scheme without diffusion. The characteristic equation reads

$$\omega^2 + (A^2 + 1 - B)\omega + A^2 = 0 \tag{9.27}$$

We keep the input of A into the reactor constant but gradually change the value of B. For the Brusselator scheme, $\Delta = A^2$ is always positive. However $T = -A^2 - 1 + B$, and

$$\rho = (A^2 + 1 - B)^2 - 4A^2 = \{(A + 1)^2 - B\}\{(A - 1)^2 - B\}$$

is a function of B. If $B < (A - 1)^2$ then $\rho > 0$ is positive and T is negative. Thus, from Fig. 9.13 we see that the singular point is a stable node. Therefore, the uniform stationary state $X^s = A$ and $Y^s = B/A$ is stable. If disturbed, the system automatically returns to these values.

If we increase B such that $(A - 1)^2 < B < (A + 1)^2$, then ρ becomes negative; the two roots are imaginary and complex conjugates. In this range of values of B, as long as $B < A^2 + 1$, T is negative and the singular point is a stable focus. The fluctuations in the system die out eventually, as the representative point of the system tends again toward the stable uniform steady state.

As B is increased further, at $B = A^2 + 1$, $T = 0$, the roots are purely imaginary and the singular point is a center. The chemical reaction has reached a critical point of marginal stability. Beyond this value of B_c, the uniform state becomes unstable. Indeed, for values of $B > A^2 + 1$, $T > 0$, and the singular point is an unstable focus. Perturbations take the system away from the homogeneous steady state.

In the Presence of Diffusion. As we have seen already, in this case the characteristic equation reads

$$\omega_m^2 + (\beta_m - \alpha_m)\omega_m + A^2 B - \alpha_m \beta_m = 0 \qquad (9.28)$$

with

$$\alpha_m = B - 1 - D_X \frac{m^2 \pi^2}{l^2}$$

$$\beta_m = A^2 + D_Y \frac{m^2 \pi^2}{l^2}$$

This characteristic equation is analyzed exactly in the same manner as before. We can see at once that if we put $\delta_m = 1 + (D_X - D_Y)(m^2\pi^2/l^2) > 0$ then $\rho = \{B - (A + \sqrt{\delta_m})^2\}\{B - (A - \sqrt{\delta_m})^2\}$. If the value of B is such that $(A - \sqrt{\delta_m})^2 < B < (A + \sqrt{\delta_m})^2$ then ρ is negative and the two roots are complex conjugates. Moreover, if $B \geqslant A^2 + 1 + (D_X + D_Y)(m^2\pi^2/l^2)$ one sees that the reference uniform steady state is an unstable focus and the fluctuations are amplified in an oscillating fashion.

However, in the presence of diffusion a new instability appears that was not present in the nondiffusing example. One may verify by direct insertion that if

$$B \geqslant 1 + A^2 \frac{D_X}{D_Y} + \frac{A^2 l^2}{D_Y m^2 \pi^2} + D_X \frac{m^2 \pi^2}{l^2} \qquad (9.29)$$

then the steady state is unstable as the singular point is a saddle point.

By an appropriate choice of various parameters, most singular points listed in Fig. 9.13 may be shown to exist for the diffusion-governed Brusselator. In what follows, we shall only consider two main singular points, an unstable focus and a saddle point.

9.8. BEYOND UNSTABLE POINTS

In the preceding section we showed that if one linearizes the nonlinear rate equations of the Brusselator scheme around a given steady state, the stability of the latter may be deduced immediately from Fig. 9.13. A linearized equation only describes the immediate vicinity of the steady state under consideration, and thus the information it contains is only valid in this region. In particular, the phase trajectories and their time behavior are limited to a small region around the steady state, as shown in Fig. 9.3.

For example, if it is found that the steady state is stable toward infinitesimal perturbations, there is no guarantee that it will remain stable if perturbed strongly. Moreover, a given instability does not guarantee the onset of a specific type of spatio-temporal behavior, nor does it supply much information as to where the system goes after leaving a steady state. However, if for a given $B = B_c$ the thermodynamic branch becomes unstable, the onset of an organized state is probable. Nevertheless, the explicit nature of the new states must be determined from the full nonlinear equations. Two types of procedure may be used for the construction of the bifurcating solutions:

1. Once the parameters of the unstable region are determined from linear stability analysis, the full nonlinear equations may be integrated numerically.
2. If an unstable steady state has been detected, an approximate analytical solution may be constructed for the bifurcating branch around this point.

The second procedure is part of an active field of mathematics which studies nonlinear differential equations. The analytical methods of bifurcation analysis are involved, and are beyond the scope of the present book. Thus, in what follows we shall be only concerned with the numerical solutions of the Brusselator scheme.

9.8.1. Brusselator as a Chemical Clock

We have seen that the Brusselator scheme in the absence and also in the presence of diffusion processes exhibits an unstable focus. Such an instability may arise for $B_c > A^2 + 1$ and $B_c \geqslant A^2 + 1 + (D_X + D_Y)(m^2\pi^2/l^2)$. With these parameters the small fluctuations and thus the total solution $X = X^s + x(t)$ and $Y = Y^s + y(t)$ leave the immediate neighborhood of the steady state and may reach a new stable regime.

A direct numerical integration of the rate equations (9.4) reveals a regular time periodicity of $X(t)$ and $Y(t)$, as shown in Fig. 9.14. With constant frequency, concentrations of intermediates rise and fall regularly, thus defining an endogenous chemical clock. This behavior of the Brusselator was shown by Lefever and Nicolis. It mimics the experimentally observed chemical oscillations in the BZ reaction.

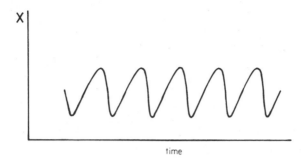

Fig. 9.14.   In the Brusselator model for $B > B_c$, the concentration of X oscillates periodically in time.

The phase trajectories of the motion may be constructed from the rate equations (9.4):

$$\frac{dX}{dY} = \frac{A - (B+1)X + X^2 Y}{BX - X^2 Y} \tag{9.30}$$

$Y = Y(X)$ computed from this equation is represented in Fig. 9.15a. It shows that no matter where they initiate, all the trajectories of the phase plane finally converge toward a unique closed curve which surrounds the singular point and is called a *limit cycle*.

Again, a convenient way of visualizing the motion is to imagine that the representative point of the system, the small ball, is constrained to move over an imaginary surface that may be constructed in the following manner. A trumpet and a tuba are fitted together in the manner of Fig. 9.15b. The heavy lines show the surfaces that may contain representative points of the system. Whatever the initial position of the ball, it invariably will fall and roll around the intersection of the two instruments. This trajectory is the mechanical analog of a limit cycle.

9.8.2. Concentration Waves in the Brusselator

Let us imagine a thin layer of unstirred Brusselator reagents in a petri dish. The reaction boundary is a circle of a well-defined radius R. As matter must not leave the petri dish, the differential equations (9.6) must be solved subject to zero-flux boundary conditions.

In the presence of diffusion, as one increases the value of the bifurcation parameter B, several distinct singular points appear. The change in solutions

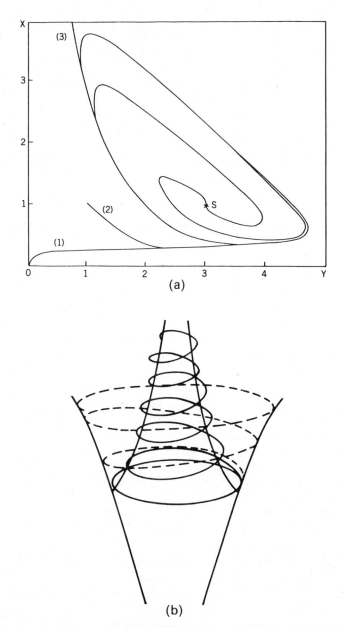

Fig. 9.15. The phase plane trajectory of the Brusselator for $B > B^c$ is a stable limit cycle. All trajectories are attracted toward a closed curve surrounding the steady state.

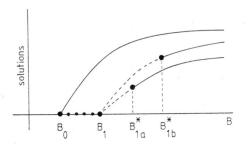

Fig. 9.16. The bifurcation diagram of the Brusselator model.

as function of B is depicted in Fig. 9.16, which we call a *bifurcation diagram*.

Below B_0 the system follows the thermodynamic branch. Beyond this point the stable solution is a limit cycle. Had we started beyond B_1 the unstable thermodynamic branch would have bifurcated into two new but unstable solutions.

In 1975 Erneux and Herschkowitz-Kaufman investigated the nature of the bifurcating solutions beyond these two points, by analytical methods as well as by computer simulations. They found that at B_{1a}, one of these branches

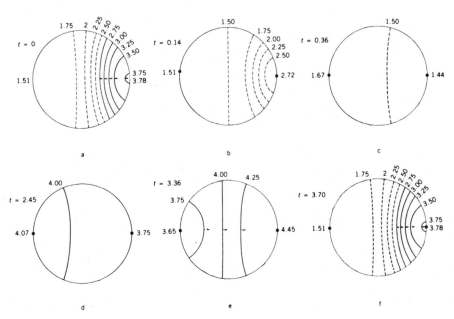

Fig. 9.17. Brusselator reagents in a petri dish are subject to zero-flux boundary conditions. Full and broken isoconcentration curves refer respectively to concentrations larger or smaller than the unstable steady state. Parts (a) to (b) represent patterns of various stages of the periodic solution.

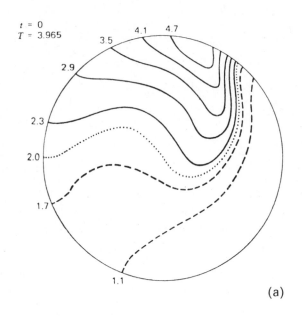

(a)

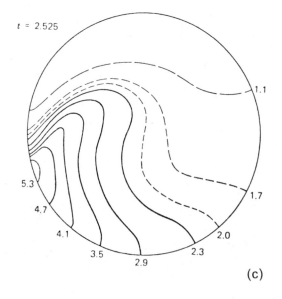

(c)

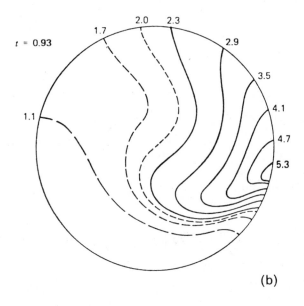

(b)

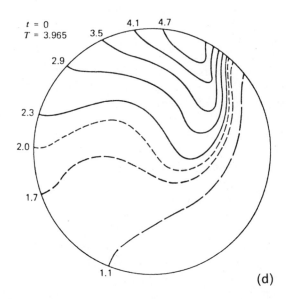

(d)

Fig. 9.18. If the value of B is increased in the experiment of Fig. 9.17, a rotating wave appears and moves periodically. The four illustrations show the time evolution of the wave.

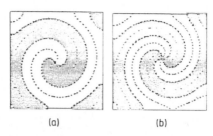

(a)   (b)

Fig. 9.19. One-armed and two-armed spiral waves may be computed in a model scheme. (Courtesy of S. Koga, 1982, reprinted by permission Prog. Theor. Phys.)

becomes unstable and the system may choose a new regime. Concentration waves travel in the system in periodic fashion and appear as outgoing quasi-parallel isoconcentration curves (see Fig. 9.17).

Beyond the bifurcation point B_{1b}, the isoconcentration curves rotate in time in the petri dish and come back to their initial state after a given time interval (one period). The periodic phenomenon continues as long as the same constraints are held in the system. If, for example, the highest isoconcentration curve is associated with blue color, this may correspond to a rotating wave in the petri dish (see Fig. 9.18).

Although these results show unambiguously the presence of spatio-temporal periodicity, the wave-like solutions of Fig. 9.18 produced by the Brusselator reaction, do not completely account for the spiral or concentric waves seen experimentally in the BZ reaction.

Tyson and Fife have constructed a two-variable reaction-diffusion system based on the Oregonator scheme of Section 9.2 and showed, using analytical methods, the presence of concentric propagating waves. The mathematics of the model is too involved to be reproduced here. Figure 9.19 shows one-armed and two-armed spirals obtained from computer simulations by S. Koga.

9.8.3. Spontaneous Emergence of Spatial Patterns

We have seen that an instability of the saddle type may also arise in the diffusion-governed Brusselator. Analytical and numerical computations performed by G. Nicolis, J. Auchmuty, R. Lefever, and M. Herschkowitz-Kaufman have shown the emergence of a wide variety of time-independent spatial patterns beyond the unstable singular points. The nature of the patterns is determined by the boundary conditions imposed on the system, the geometry and the size of the reactor, as well as by the various diffusion coefficients of the different molecular species.

If the input of B is adjusted such that $B < B_c$, the concentrations of X and Y in the reactor remain homogeneous. However, if $B > B_c$, the homogeneous

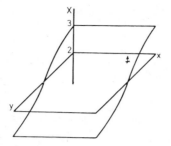

Fig. 9.20. In a square box subject to zero-flux boundary conditions, a nonhomogeneous concentration profile, or a polar structure, can arise spontaneously.

state is potentially unstable and any small fluctuation brings the system into a new stable state which is necessarily nonhomogeneous. Figure 9.20 shows a pattern obtained in an unstirred square reactor of size *l*, subject to zero-flux boundary conditions and filled with a thin layer of Brusselator mixture. We see the onset of a gradient in the concentration of *X* and *Y* in the system. At one end, the concentration of *X* is above and at the opposite end below its value in the homogeneous state. The reverse is seen for the *Y* concentration. We thus

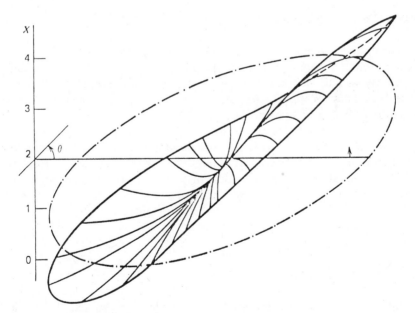

Fig. 9.21. A two-dimensional, polar, steady-state dissipative structure in a circular geometry. Broken dotted lines show the uniform steady state.

speak of a spontaneous generation of *polarity* in the medium. As B increases further, new and complex patterns appear in the reactor.

Figure 9.21 represents a two-dimensional reactor of circular geometry. Again, for an appropriate B_c, a polar structure appears in the medium. However, in this case, in the same structure a region of space around the point $\theta \simeq 0$ is highly organized, whereas another quasi-uniform region is seen around $\theta \simeq \pi/2$.

The coefficients of the characteristic equation (9.28) are functions of the geometrical dimension l of the reactor. If we keep B below the critical value B_c and increase the size of the reactor gradually, for a critical length l_c a new saddle-type instability appears. For zero-flux boundary conditions, the first pattern is again similar to Fig. 9.20; however, as l increases, patterns with several maxima may appear.

Thus we may conclude that the presence of saddle-type instabilities in a reaction-diffusion system leads the homogeneous steady states into *symmetry-breaking instabilities*. The ensuing non-homogeneous patterns are the chemical analog of Benard rolls in hydrodynamics.

Although stable time-independent spatial patterns have been shown to exist for a wide class of chemical schemes, such patterns have not yet been observed unambiguously in any real experimental system.

9.8.4. Chaos in the Brusselator

Sometimes the solutions of the nonlinear deterministic differential equations exhibit random behavior which is the property of the macroscopic system. Recently, with the help of numerical simulations Kuramoto showed that, if in the Brusselator $A = 2$, $B = 6.5$, $D_X = 1$, and $D_Y = 0$, the reaction-diffusion equations (9.6) exhibit chaotic behavior. With these parameters it is easy to show that the homogeneous steady state of the Brusselator is indeed an unstable focus. Fig. 9.22 shows such a chaotic spatio-temporal pattern in the concentration of intermediate X, obtained by direct integration of reaction-diffusion equations. A cross section of these patterns with a plane perpendicular to the space axis reveals a pattern similar to that of Fig. 9.22, which should be compared with the chaotic behavior observed in the BZ reaction shown in Fig. 8.12.

*   *

*

Simple as it may look, the Brusselator exhibits extremely complex bifurcating solutions resulting from the instability of the thermodynamic branch of solutions. However, multiple homogeneous steady states have not been

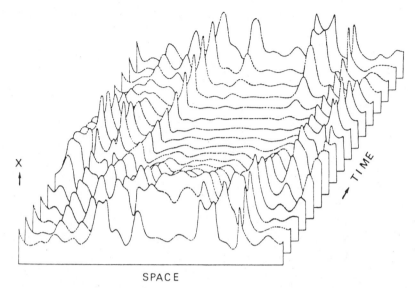

X
↑

SPACE

Fig. 9.22. Chaotic spatio-temporal patterns in a Brusselator model. (From Kuramoto, 1981, reprinted by permission of North-Holland Physics Publishing.)

observed in this scheme. These types of bifurcations are important for the understanding of hysteresis, and the all-or-none type of behavior of biological systems. Therefore, we now turn to a simple, one-variable model which exhibits multiple steady-state solutions.

9.9. A SIMPLE MODEL WITH MEMORY

We have seen that the simplest possible oscillating reaction, namely the Brusselator scheme, already models almost all the peculiar behavior of the BZ reaction. However, the steady-state values of the Brusselator, $X^s = A$, $Y^s = B/A$, are single-valued constants and are determined by the input flows. Therefore, this scheme cannot model the multiple steady-state behavior or hysteresis effect of the BZ reaction (see Section 8.7). However, the one-variable simple Schlögl model, Eq. (6.14), exhibits multiple steady states.

We have already shown that, far from thermodynamic equilibrium, the homogeneous steady state of this model may become unstable. The bifurcating solutions of the model may be found analytically. The time change of the intermediate X is given by

$$\frac{dX}{dt} = -k_{-1}X^3 + k_1 BX^2 - k_2 X + k_{-2} C \qquad (9.31)$$

Moreover, as we are only interested in finding multiple steady states, Eq. (9.31) reduces to an algebraic equation of the form

$$X^3 - aX^2 + bX - c = 0 \qquad (9.32)$$

Although the formal solution of this equation may be found in elementary books on algebra, it is not easy to visualize the dependency of the solutions on bifurcation parameters.

Faced with a cumbersome or intractable problem, mathematicians define new variables without prejudice to the nature of the process at hand. With the help of a new variable, an impossible problem may become solvable, and a complex one may take a more simple form. For our model, following G. Nicolis and J. W. Turner we define a new variable $x = (X/B) - 1$ and two new parameters $\delta = k_2 - 3$ and $\delta' = (C/B) - 1$. Then we choose $k_1 = 3/B^2$, $k_{-1} = 1/B^2$, and $k_{-2} = 1$. Introducing these quantities into Eq. (9.32) we get

$$-x^3 - \delta x + (\delta' - \delta) = 0 \qquad (9.33)$$

The nature of the solutions of this algebraic equation is crucially dependent on the numerical values of δ and δ'. Moreover, we must bear in mind that X is a concentration of a chemical and therefore may only take real and positive values. Consequently, although a third-degree algebraic equation always has three roots, we must consider only the real values of x satisfying the condition $X/B = x + 1 > 0$. The imaginary values are irrelevant.

Let us choose $k_2 < 3$ and start progressing from the equilibrium state by increasing gradually the concentration of B. At equilibrium, $B = C/3k$ and x has the unique value of $x = 2$, whereas $X = 3B$. As B increases, x changes but remains single-valued till the value of $B_c = C/(k_2 - 2)$ is reached; then $\delta' - \delta$ vanishes and the polynomial Eq. (9.33) takes the simple form

$$-x^3 - \delta x = -x(\delta + x^2) = 0$$

As $\delta = k_2 - 3$ is assumed to be negative, the three real roots of the polynomial are computed immediately and are found to be $x_1 = 0, x_2 = +\sqrt{-\delta}$, and $x_3 = -\sqrt{-\delta}$. Fig. 9.23a shows the three solutions of x as a function of the constraint parameter δ and represents the bifurcation diagram of the model. The point 0 from which the three solutions emerge is the bifurcation point. If we go back to the old variables, we find three concentrations for X, namely $X_1 = B, X_2 = B(1 + \sqrt{3 - k_2})$, and $X_3 = B(1 - \sqrt{3 - k_2})$. Therefore, B_c is a bifurcation point, from which the thermodynamic branch bifurcates into three simultaneous solutions. If B increases further, the system enters a domain

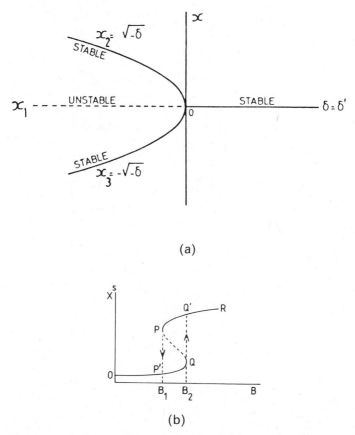

(a)

(b)

Fig. 9.23. (a) Bifurcation diagram for the Schlögl model. (b) Steady-state concentration X^s versus a bifurcation parameter. Broken lines denote the unstable branch. Between B_1 and B_2 the system shows multiple steady states and hysteresis. If B increases, X^s changes along OP'QQ'R and returns along the path RQ'PP'O.

where only one solution exists. Fig. 9.23b reports the change in X as a function of the bifurcation parameter B.

It is important to notice that multiple steady states appear only in a well-defined and usually very narrow range of parameters. In the present case, bifurcation appears for all combinations of parameters satisfying the condition $\delta = \delta'$ or, equivalently, $(C/B) - 1 = k_2 - 3$. This fact illustrates the dangers of numerical integration of nonlinear differential equations when searching for bifurcating solutions. They can be missed very easily, as even with the most powerful computers it is not possible to scan every combination of parameters of the system. In the absence of any analytical clue delimiting the parameter values, the numerical search for bifurcating solutions is similar to looking for a needle in a haystack.

Fortunately, with the help of linear stability analysis or thermodynamic stability criteria, the range of parameters for which the system may become unstable may often be determined. As bifurcations are intimately associated with instabilities, whenever a thermodynamic branch becomes unstable it bifurcates into a new stable branch.

It is interesting to note that De Kepper and Boissonade showed that if X is taken as the concentration of the key intermediate $HBrO_2$ in the Oregonator model of the BZ reaction, at steady state the value of X is a solution of an equation similar to Eq. (9.32).

The concepts of stability, bifurcation, and multiple steady states may be illustrated by the following example borrowed from classical mechanics. We take a port wine bottle with a profile as depicted in Fig. 9.24a and hang it from the neck. We introduce with great care a small ball into the bottle, such that it comes to rest at point 0. We know by experience that a slight knock or disturbance to the bottle will topple the ball into either one of two positions, C or D. The three positions 0, C, and D are all available to the ball; however, 0 is an unstable equilibrium point, while C and D are stable.

Now we heat the bottom of the bottle to the melting temperature of the glass. Under the action of gravity and heat, the bottle takes a new shape, as shown in Fig. 9.24c. The ball rolls into the bottom of the bottle and remains there in a stable equilibrium. With this new shape, only point 0 is stable.

Imagine that in the process of transition from shape (a) to shape (c) we were able to freeze the bottom of the bottle into a completely flat surface, as shown

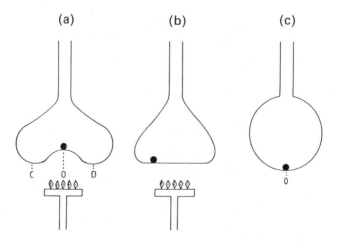

Fig. 9.24. A visualization of the concept of stability and multiple steady states.

in Fig. 9.24*b*. Now the ball does not have any preferential position; it may be moved from 0 to any other equilibrium point. All points of the flat bottom are in *marginally stable equilibrium*.

9.10. SECONDARY BIFURCATIONS

The bifurcation diagram of Fig. 9.23*a* is called *supercritical*, since new branches appear beyond the bifurcation point. However, this is not always the case; a bifurcation diagram may look like the one depicted in Fig. 9.25*a*. Beyond λ', three simultaneous solutions exist. However they are disconnected from each other; only by increasing λ to λ_c, do we reach a critical point where the two branches join together at the bifurcation point.

In some instances, the bifurcation diagram may take the form of Fig. 9.25*b*. Beyond $\lambda = \lambda'$ two new stable branches appear along with two unstable branches; however, the steady homogeneous state remains stable beyond this point until the value $\lambda = \lambda_c$ is reached. In the vicinity of this point, the thermodynamic branch is unstable and two stable solutions remain.

Our jogger in Fig. 9.2, after the first bifurcating path, may come across a new bifurcation point which branches from the already bifurcated path he has been forced to take. Similar situations may appear in chemical reactions. If one increases the constraints on a system, a newly bifurcating solution may become unstable and give way to another, more stable, new solution. The process may continue several times, as shown in Fig. 9.26. This diagram exhibits *primary*, *secondary*, and *tertiary* bifurcation points.

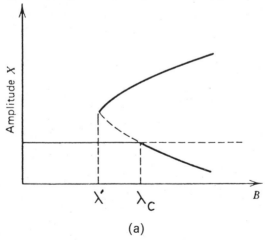

(a)

Fig. 9.25. (*a*) Transcritical bifurcation diagram. (*b*) Subcritical bifurcation diagram. Full and broken lines represent, respectively, stable and unstable solutions.

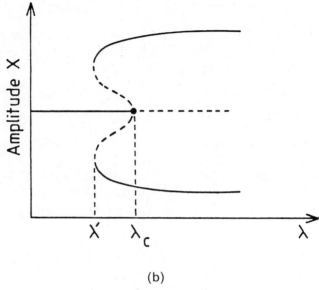

(b)

Fig. 9.25 (*Continued*)

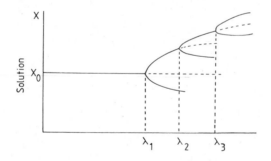

Fig. 9.26. A bifurcation diagram showing primary, secondary, and tertiary bifurcations.

CONCLUSION TO PART II

In this part we have summarized an approach that has completely changed our view of the chemical sciences. Modern chemists have discovered that reactions are not always predictable and processes do not always evolve smoothly. However, in order to see this, they must substitute flow reactors for the more usual chemical laboratory equipment. Moreover, unless appropriate feedback processes are present, chemical reactions cannot exhibit the peculiar behavior discussed in Chapter 8. The field of the new chemistry is in its infancy. Certainly, in the near future, many more self-organizing chemical reactions and new exciting properties will be discovered.

We should, however, stress again that the new chemistry reported here is a recent discovery, and at the present time is not even part of the curriculum of most high schools or institutions of higher learning.

The development of the new chemistry has coincided with vast progress in biochemistry; and as the molecular details of biological processes have become known, mathematical models have been constructed. The third and final part of this book gives some examples in which the methods of the present section have been used to elucidate several important aspects of self-organization in living organisms.

Part III

SELF-
ORGANIZATION
AND COHERENCE
IN BIOSYSTEMS

What is life?

E. Schrödinger 1945

INTRODUCTION TO PART III

The spontaneous onset of self-organized states in chemical reactions, when there are no thermal effects, is a recent discovery. Such phenomena have completely changed our view of the physical sciences and their relation to the biosphere. Bulk matter is no longer an inert object that can only change if acted upon. On the contrary, it can have its own will and versatility and ability for internal organization. Such findings have considerably narrowed the wide gap that existed between matter and life. We now have enough elements in hand to allow us to hope for a distant day when matter can be made to become alive through the action of ordinary physico-chemical laws.

Such a perspective is a revolution in science. For centuries, the biological and physical sciences traveled along divergent roads, each following its own laws. To see this more clearly, let us go back again to the nineteenth century.

In addition to the idea of entropy, the science of the nineteenth century was dominated by a second evolutionary concept. Darwin claimed that the biosphere was not created as such. The more simple species evolved in time into many different forms, and today's species were selected by the environment because of their survival value.

Darwin's ideas implied that simple organisms were endowed with a deterministic developmental program, and were subject to random and unplanned errors which, accumulating in individuals, gradually produced a new species. The individual organisms in these species were selected or rejected by the environmental constraints. In such a process, more and more new and increasingly complex species appeared, which were able to survive insofar as they were in harmony with the ecological constraints. Thus the Darwinian theory of evolution introduced a historical dimension into the taxonomy of species.

Today we think that the birth of man, certainly one of the most complex species of the biosphere, may be traced back to 10^6 years ago. His ancestors, the unicellular bacteria, appeared some 3.5 billion years ago.

Obviously, the two evolutionary laws of the nineteenth century postulated two totally opposite tendencies which introduced a divorce between matter and living species. How could one reconcile one evolutionary law which predicted a maximum disorder, simplicity, stillness, permanence, and the same fate for all with another evolutionary law which described complexity, randomness, and diversity. Under such circumstances, a break between physico-chemical systems and living matter was inevitable. Matter and man became irreconcilable. The biological sciences had to assume their own vitalistic forces, governed by their own laws. These were completely outside Newtonian mechanics and equilibrium thermodynamics or any other macroscopic law of nature.

The gap between the biological and the physical sciences was narrowed only in the mid-1960s. Oddly enough, the coming together of biology and physics resulted from two apparently antagonistic developments. In the late nineteenth century it became obvious that the unit of all living species is the cell—a bag full of various gigantic organic molecules that could even be synthesized in laboratories. The extraordinary discoveries made in molecular biology in the last decades have identified a great number of these molecules.

At the same time it became obvious that nonequilibrium conditions are a universal necessity for all life. A cell can survive only if it is subject to the flow of nutrients. Biochemical investigations showed that nonlinear kinetics, in the form of positive and negative feedbacks, are practically the rule in biochemical processes, and chance events or "mutations" are still witnessed today in simple organisms, such as bacteria. Moreover, a cell is not just a bag of molecules; it is a highly structured entity with well-defined temporal and spatial organization. The proper understanding of its behavior was beyond the reductionist methods of molecular biology.

The concept of dissipative structures went in an opposite direction by showing that, under appropriate conditions, inert bulk matter is no longer simple and may show a great variety of complex behaviors reminiscent of living states.

If we parallel the properties and conditions for the onset of self-organized structures with the important elements in the history of biological evolution, we come to the conclusion that both seem to obey the same laws and exhibit similar behavior. Therefore, it is tempting to try to throw a bridge between the living cell made of spatially and temporally organized biomolecules and chemical systems which organize themselves and become ever more complex under nonequilibrium constraints. In such a perspective, the phenomenon of life seems "natural," an inevitable consequence of the ordinary physico-chemical laws which govern the entire universe.

The bold and audacious hypothesis which assumes that life has been created as a result of the self-organization of matter is new. At the present time it seems the only valid hypothesis which reconciles matter and life. Ultimately, such an idea must be confirmed in laboratory experiments. We are at the very beginning of such an endeavor, and the road from molecule to life is still very long and full of pitfalls. However, we are entitled to hope that sometime in the future it can be proved unambiguously that self-organized properties of reacting and flowing systems constitute the missing link in the evolution of molecule to man.

We are encouraged in such an expectation by the existing theoretical models which show the feasibility of such an endeavor. Indeed, relying on the self-organizational properties of matter subject to nonequilibrium constraints, M. Eigen has elaborated a plausible scenario in which a primitive cell emerges from the interaction of proteins and nucleic acids. The problem of the synthesis of these molecules from C, H_2, N_2, and H_2O has partially been resolved.

The progress of molecular biology in the last decades has been phenomenal. The structural details of cells have been photographed with the help of electron microscopes. The biochemical mechanisms of quite a few important biological processes have been elucidated. The amino-acid sequence of some proteins has been determined. This progress gains momentum at even faster speed. However, life in its endless diversity offers many mysteries that these spectacular advances have not, and we would like to say cannot, entirely elucidate. A reductionist view of the living state cannot explain a cooperative behavior at the macroscopic level.

Life is but rhythm, pattern, cooperation, coherence, and change. The information necessary for all biological processes is conserved and reproduced as a new organism. Nature is full of biochemical timers; heartbeats are the most familiar to us. There are also circadian rhythms and biochemical clocks. The emergence of living organisms, endowed with well-defined and complex patterns, from a tiny seed or a small egg has been a long-standing puzzle for the biologists. We must still find an answer to the question of how in the embryo matter organizes itself and unfolds in time into patterns of function and form.

A simple nerve cell has the capacity to generate and propagate electrical current. The cooperation of billions of such cells generates action, feelings, and thought. These manifestations are the properties of the brain, not of a single neuron.

Direct observation and experimentation with biological materials, such as cell extracts—called *in vitro systems*—or with intact cells and organisms—called *in vivo systems*—reveal the existence of dissipative structures at all levels of biological organization. Today we feel that a reductionist view of biology is the first and a crucial step in the scientific investigation of the processes of life. However, it must be incorporated in a dynamic framework in order to account in a satisfactory manner for those aspects of biological systems which discriminate between life and an assembly of extremely sophisticated molecules.

A dynamic approach similar to the one used in Part II is only applicable to biological systems if the molecular details of the mechanisms underlying the global properties of these systems have been elucidated by classical methods of molecular biology. A great number of theoretical models have been constructed in order to try to account for the self-organized states observed in various biological processes.

In this last part of the book we intend to give some illustrative examples and to indicate how they can be understood in the context of the theory outlined in the preceding chapters.

We would like first to address a warning to the reader with biological training. He may find that some of the models are based on old data, or incomplete information, or simply on experiments he personally does not believe in. This must not lessen his interest in the models, as their main purpose is to illustrate a dynamic approach to a given problem. They show how to bring together important biological information with a minimum number of equations. They also indicate how to extract global properties from such simplified equations. Following this procedure, the reader will be able to construct his own "better" models, since a large class of models may show qualitatively the same properties.

The elementary notions of biochemistry, necessary for the understanding of the concepts involved in each specific model, will be of some help to the general reader.

The examples treated in this part have been selected according to the following criteria. Each example illustrates the role of dissipative structures in the organization of life at a distinct stage of the biological processes, and each example is parallel to one of the self-organizing properties of the BZ reaction, and is treated by the mathematical methods of Part II.

Chapter 10 is devoted to the fascinating problem of the origin of life. After introducing the necessary notions of biochemistry and prebiotic evolution, we show the role of multiple steady-state-type bifurcations in the birth of life. We give plausible scenarios as to how nonlinear dynamics and nonequilibrium constraints may have produced a primitive cell.

Dissipative structures at the cellular level are the subject of Chapter 11. Cells may show rhythmic behavior in response to external or internal signals. Yeast cells transform sugar into alcohol via a long chain of biochemical reactions. Regular and periodic change in the concentration of the intermediate products can be seen in appropriate experimental conditions. We show that such a biochemical clock may be compared to the chemical clock seen in the BZ reaction.

Unicellular organisms also use multiple steady-state-type bifurcations in order to cope with environmental constraints. Under certain conditions the same mechanism may give rise to a biochemical clock.

Unicellular organisms may communicate by sending chemical signals that cover a large territory. Perhaps such events can be understood with the help of the properties of propagating chemical waves, as seen in the BZ reaction. In

Chapter 12 we construct a model for the first stages of aggregating colonies of slime molds.

Multicellular organisms are characterized by spatial order. The zebra's stripes, the leopard's spots, and the spatial arrangements of the organs of our own bodies are but a few examples of such morphological order. We think that a biochemical pattern underlies all these apparent organizations. In Chapter 13 we relate these biological patterns to the spontaneous emergence of spatial organization in reaction diffusion systems.

Our brain is a masterpiece of cooperation and coherence among billions of cells. The self-organization of these cells is responsible for all activities of the central nervous system. The study of such problems in the framework of dissipative structures raises new mathematical problems. In Chapter 14 we extend the methods of Part II to these multiple-unit multiply-connected systems and show that the epileptic seizure, a pathological activity of the brain, can be understood by tracing an approach similar to the one we used for describing chemical and biochemical clocks.

These examples may be called classical; most of them were developed in the 1970s at a time of renewed interest in the mathematical approach to the self-organization of biological systems. However, the study of self-organization in relation to various aspects of animate nature continues at an ever-increasing pace. Unexpected properties and new domains of application are being investigated by a large community of scientists.

Chapter 10

THE ORIGIN OF LIFE

10.1. INTRODUCTION

In the ancient civilizations of Babylon, Egypt, India, and China it was believed that life was generated spontaneously from inanimate matter. The daily observation of nature confirmed this belief. For example, worms could be seen to develop from the decomposition of organic materials, and flies appeared spontaneously in a piece of meat left in the sun. Aristotle, who was also a great observer of nature, concluded that living organisms are generated either spontaneously or from a parent. According to him a new organism emerges from the cooperation of two "principles"; a passive principle (matter) and an active principle (form).

In the western world, Aristotle's theories were accepted and endorsed by the religious authorities, and were not questioned until the seventeenth century. At the end of this period Francesco Redi (1626–1698) advanced a new doctrine. For him, in the beginning God created all existing plants and animals; thus, life could be generated only from a preexisting life. He substantiated his ideas with the following experiment. An open jar, containing rotting fish, after a while generated flies, while an identical jar covered with a fine net remained sterile. This experiment demonstrated that the flies came from eggs and were not grown spontaneously from the rotten flesh.

In the decades that followed Redi's experiment, the invention of microscope by Antoine van Leeuwenhok (1632–1723) revealed a new kind of life: the kingdom of the infinitely small, populated with microorganisms of various sizes and shapes. For example, bacterial populations appeared spontaneously in any soup that was left aside for some time. The discovery of microorganisms revived the belief in the spontaneous generation of life. In the century that followed, biologists were divided between two schools of thought: the advocates of spontaneous generation and those who believed that life came only from life.

In 1862, Louis Pasteur showed unambiguously that life could only come from life. He demonstrated that microorganisms lived in the ambient air and finally contaminated the most sterile culture medium. In his landmark experiment, Pasteur showed that if a sterile culture broth was sealed off from the air, no microorganisms appeared spontaneously in the liquid. Pasteur's work demonstrated that even microorganisms came from parents. Therefore, living organisms appeared as singularities in nature, and it seemed that there was no connection between the substances they were made of and the inanimate matter that surrounded them. In 1859, while the controversy between the partisans of spontaneous generation and their adversaries was still strong, Charles Darwin published *The Origin of Species*, one of the most significant contribution to our scientific knowledge.

Darwin observed that variations exist among a given species. And he found in the fossil record, plants and animals different from present-day species. He thus came to the conclusion that all species came from a common ancestry. The present-day ecosystem is the result of a long evolutionary process that took a few billion years. Darwin's theory of evolution was based on three key concepts. The individual members of a species proliferated by *multiplication*; *variations* in these individuals appeared during this process. As the various living units competed for the available resources, the fittest individuals

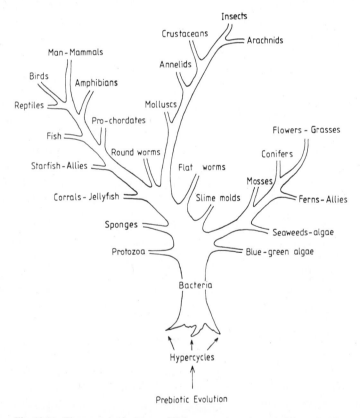

Fig. 10.1. The genealogical tree of living organisms is rooted in molecules.

survived through *natural selection*. In this process local populations diverged due to the accumulation of successful variations, and thus gave rise to new species.

This scheme implied that the ancestor of all living things was a simple, unicellular organism or a bacterium. Today data gathered from various sources corroborate Darwin's hypotheses, and may be arranged in a genealogical tree as shown in Fig. 10.1.

Till the last years of the nineteenth and the beginning of the twentieth century, there was no satisfactory answer for a scientifically minded researcher to the question of where the "seed" for the Darwinian genealogical tree came from. If unicellular organisms cannot be generated spontaneously from matter, then it was thought that perhaps they arose from a very slow evolution of matter in which simple molecules gave rise to complex biological structures. In 1913, following this line of thought, the word *chemical evolution* was introduced by Moore in the context of the origin of life. Between 1920 and 1930, a great amount of data was accumulated from various fields, preparing grounds for new assumptions concerning the emergence of the living cell from inanimate matter. Teilhard de Chardin (1881–1955) advanced the hypothesis that living as well as inanimate matter gradually organizes itself from simple elements into more and more complex entities. Thus the Darwinian tree had to be rooted in the inanimate world of the periodic table (Fig. 1.7). The first living cell emerged from these atoms after a long evolutionary process.

Our first task must be the identification of those molecules that enter the composition of living structures. We must then follow the evolutionary path that generated them from their constituent atoms.

10.2. THE PIONEERING MOLECULES OF LIFE

Carbon atoms have a unique fate in chemistry. They may react with each other, and with hydrogen, oxygen, nitrogen, and a few other atoms in order to form long chains of hydrocarbons or pentagonal and hexagonal inter-connected rings. In nature such carbon compounds are found only in living or fossilized organisms, and therefore they were named *organic* substances. Already in the late eighteenth century a few *immediate principles*, such as urea, oxalic, and malic acid, were extracted from living organisms. Thus for a long time it was thought that only *vital forces* present inside organisms could produce such molecules. However by the nineteenth century many new organic substances were extracted from biological materials, and quite a few artificial organic compounds were synthesized in laboratories. Today the organic molecules of biological origin are but a small fraction of all available carbon compounds. Their study constitutes the subject matter of *biochemistry*.

Molecules of life may be divided broadly into four categories: proteins, nucleic acids, carbohydrates, and lipids. Currently it is believed that the last two categories of molecules are latecomers in the evolutionary process, and that life originated from inanimate matter after the emergence of proteins and nucleic acids.

10.2.1. Proteins

Proteins are formed from a large number of *amino acids* that are linked together by a *peptide bond*. Amino acids are relatively large molecules that are characterized by the simultaneous presence of NH_3^+ and $C\underset{\diagdown O}{\overset{\nearrow O}{}}$-groups, and are of the general form $H_3^+N - HCR \!-\!\!-\!\!- C\underset{\diagdown O}{\overset{\nearrow O}{}}$. Here R is a hydrocarbon chain that may take a variety of forms. Only 24 different amino acids enter into the composition of all known proteins from bacteria to man. $R = H$ for the simplest of all biological amino acids, glycine, whereas R is formed of two cyclic molecules in the most complex amino acid, tryptophane. Peptide bonds are formed between two amino acids by the elimination of one molecule of water, as shown in Fig. 10.2.

Each protein has a unique amino acid sequence, called its *primary structure*, determined by hereditary factors. A typical protein may contain as many as 200 amino acids. The long polymer chain coils in space into a three-dimensional structure defined as protein *conformation*. The catalytic activity of a protein is critically dependent on its conformation. A stretched protein loses its catalytic activity.

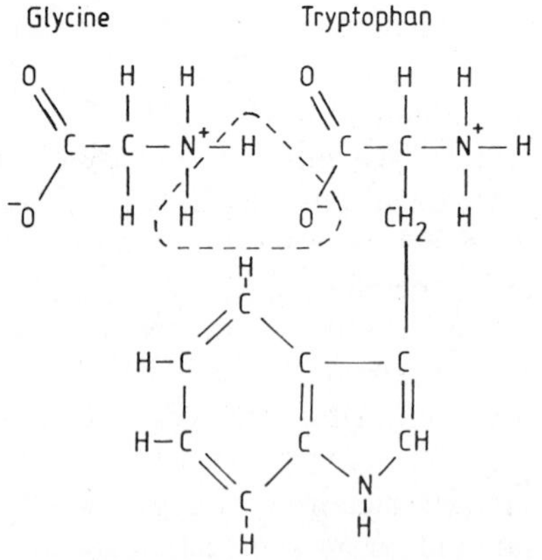

Fig. 10.2. A peptide bond is formed between two amino acids by elimination of one molecule of water.

Two amino acids of the same composition may differ in the same way as a left and right hand; they are not superimposable. For reasons not elucidated yet, nature has only chosen one hand, the left amino acids, as the building blocks for living organisms.

10.2.2. Nucleic Acids

Organisms carry a second type of giant macromolecules, called ribonucleic and desoxyribo-nucleic acids, in short RNA and DNA. The structure and function of these molecules are radically different from proteins. DNA molecules carry in them all the information and instructions necessary for the synthesis of every bit of biological material, including their own formation, thus perpetuating the species. The language used by DNA molecules is somewhat similar to the ordinary rules of linguistics. There is an alphabet with a finite number of symbols. Words are formed from this alphabet by a well-defined sequence of letters. An ordered succession of words forms a meaningful sentence. The sum total of many sentences expressing interrelated events defines such masterpieces as human beings.

The alphabet of life has only four molecules, which are divided into two distinct groups: *purines* and *pyrimidines*. The two purines, adenine and guanine, and the two pyrimidines, cytosine, and thymine, are all bases; they are shown in Fig. 10.3a together with another base uracil which only enters into the structure of RNA. It is customary to designate them simply by their initials, A, G, U, C, and T.

The ancient Babylonians preserved the order of their symbols by engraving them on clay tablets. Today these templates may still be reproduced. Similarly the *templates* of living organisms are made of a long succession of phosphate and sugar molecules forming the backbone to which the four bases are attached. Figure 10.3b shows such a gigantic DNA molecule, the first copy of which was written probably 2.5 billion years ago.

In DNA language all words or *codons* are written with three letters, and they indicate "start," "stop," or designate one of the 24 amino acids. A complete sentence or a *gene* designates a specific protein. Synonyms are very frequent in DNA language or *genetic code*. For example, CAU and CAC both mean the amino acid alanine. The DNA molecule of a given organism is the complete book of the saga of that organism. The ancient history and the future development of the organism are all imprinted on the DNA template.

If a template is not used, it must be wrapped in an envelope in order to protect it from time, erosion, and harsh environment. Organisms accomplish this by wrapping two identical DNA molecules around each other in an helicoïdal twist. All information-carrying bases are imprisoned safely inside the double helix. However, the two chains run in opposite directions such that

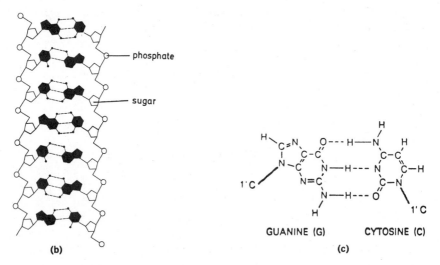

(a)

(b)   **(c)**

Fig. 10.3. (a) The five bases entering the structure of DNA and RNA. (b) Genetic information is engraved in the double-stranded DNA. The two strands are held together by means of hydrogen bonds, according to the Watson-Crick pairing rule. Adenine pairs with thymine and guanine with cytosine. (From *Biochemistry*, 2nd ed. by L. Streyer. W. H. Freeman and Company. Copyright © 1975). (c) Hydrogen bonds.

the pyrimidines form hydrogen bonds with purines of the *complementary strand*. In the process of DNA replication, the *double helix* separates into two branches, and a new complementary strand is synthesized from each branch according to the scheme of Fig. 10.4. Errors, or *mutations*, are possible during the replication process, either because of substitution, deletion, insertion, or any defect in the copying mechanism. An error-ridden DNA is called *a mutant* of the original molecule.

The discovery of DNA is a twentieth-century affair; the double helicoïdal structure of the molecule was discovered by Watson and Crick in 1956.

The most remarkable fact about the living state is that the genetic code of all

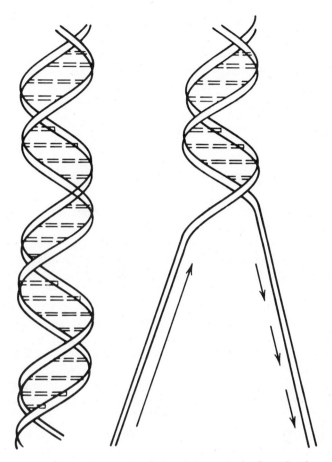

Fig. 10.4. DNA unwinds at the replication fork and the synthesis of two daughter molecules proceeds on both strands. One of the strands is synthesized in the form of fragments. The process requires the presence of several polypeptide chains (enzymes).

three and a half million species of known vegetals and one million species of animals is the same.

Bits of DNA containing one sugar, one base, and one or more phosphates are called *nucleotides*. They play a very important role in cell life as DNA precursors or enzyme complements, as we shall discuss later. One of these nucleotides, adenosine triphosphate, ATP, has a primordial role in cell metabolism. As we shall see, this molecule is the energy currency in all living species.

Living organisms use also *ribonucleic acid* or RNA, which differs from DNA on three counts: the sugar desoxyribose is replaced by another sugar, ribose;

the base thymine is replaced by uracil; and the structure is single-stranded. An RNA strand is able to fold on itself such that a pair of complementary strands form a *hydrogen bond* which stabilizes the molecular structure.

The various kinds of RNA serve as mediators for the flow of genetic information from the DNA into proteins. They alone are in contact with amino acids and proteins. Therefore, it seems natural to assume that RNA is the first informational biomolecule that evolved in the prebiological environment.

10.3. THE ADVENT OF THE FIRST BIOMOLECULES

We start the story of the origin of life from a day when the planet Earth was already formed and a few simple molecules were held in its surroundings by gravitational forces. We use the word *story* purposefully, since the theory we are about to develop may be constructed in several different versions. Data about the conditions on the surface of the primitive earth and the chemical composition of its atmosphere are fragmentary and uncertain. Every set of hypotheses about these conditions implies different biochemical pathways leading to prebiological evolution. All hypotheses start from approximately the same simple molecules, and most of them arrive at the same "happy ending": the various molecules that form the building blocks of living material in nature could have been formed very easily on the surface of the primitive earth. We shall adopt here the most popular and, historically, the first set of conclusive experimental data for the exposition of the theory of chemical evolution.

In order to account for the abiotic formation of organic materials, A. I. Oparin and, a few years later, J. B. S. Haldane independently postulated that the atmosphere of the primitive earth was very different from the present-day mixture. According to these investigators, the primitive atmosphere was composed of a mixture of hydrogen molecules, combustion gas methane, CH_4, water vapor, and ammonia. Today we have good reasons to think that their hypothesis was probably correct, since one can show experimentally that in the presence of an excess H_2 (which seems to have been the case in the primitive atmosphere), carbon, oxygen, and nitrogen cannot remain free but form H_2O, NH_3, and CH_4 molecules.

10.3.1. Prebiotic Synthesis of Amino Acids

The prebiotic atmosphere was constantly bombarded with intense radiation emitted from the sun, a condition that gave rise to the synthesis of a large quantity of organic molecules. These molecules accumulated in the oceans and gave rise to a *primitive soup*. Life originated from this soup via a long evolutionary process. (In another version of the story, the first biomolecules were formed on the bare and hot rocks and were washed into the sea by rain water.)

In the early 1950s S. Miller designed an experiment in order to test the validity of the Oparin and Haldane hypothesis. He filled a chemical vessel with the presumed primitive atmospheric gases and introduced a discharge of 160,000 volts into the mixture, thus simulating the action of lightning. When the reaction mixture was analyzed carefully, Miller could find all the amino acids which are the building blocks in the construction of biological material. He also found a large number of organic compounds that are not used by nature in the edifice of life. Miller's experiment showed for the first time unambiguously that biomaterials could be synthesized spontaneously in nonbiological conditions. This approach opened a new field of investigation that took the name of *prebiotic chemistry*.

Miller's followers introduced new methods of experimentation and new sources of energy that simulated all possible forms of energy input that were available in the atmosphere of the primitive earth. M. Calvin chose nuclear energy, which could have been available in radioactive minerals, or come from space in the form of ionizing radiation. He bombarded the primitive atmosphere with a flux of high-energy electrons which were accelerated by a cyclotron. In this way he obtained amino acids, sugars, urea, fatty acids, and other important organic substances.

10.3.2. Laboratory Syntheses of Nucleotide Fragments

Cyanhydric acid, HCN, and ammonia had been detected in comets. In 1960 J. Oro combined these substances in a water solution that he heated for 24 hours at 90°C. To his great surprise he synthesized *adenine*, a molecule which is part of the genetic alphabet and a vital element in the energy exchanges of the cell. C. Ponnamperuma later synthesized adenine from a primitive atmosphere by performing the same reaction in a cyclotron. Detailed analysis of the prebiotic reaction scheme revealed that, in a first stage, CH_4 and NH_3 produced cyanhydric acid, HCN, and the combination of methane and water molecules gave rise to formaldehyde. Later, these two highly reactive molecules could combine and form the various organic molecules. The proof of these hypotheses came from two important experiments.

Ponnamperuma submitted a solution of cyanhydric acid to a battery of ultraviolet lamps, thus simulating the action of solar radiation. After a week he found in the reaction medium adenine, guanine, and urea. Oro and Ponnamperuma independently repeated the preceding experiments, this time with a formaldehyde solution, and synthesized ribose and deoxyribose, which are two sugar molecules that enter the composition of genetic material.

Hot gas and incandescent material projected from volcanoes were another source of energy available on the primitive earth. Sidney Fox simulated the prebiotic synthesis around erupting volcanoes by heating in an oven the usual prebiotic mixture of H_2O, CH_4, and NH_3 to 1000°C. He obtained all but two of the existing amino acids (sulfur enters into the composition of both of them).

By the mid-1960s it became evident that all biologically relevant organic molecules or building blocks, such as amino acids, sugars, and adenine, could be synthesized with great ease under conditions that most probably prevailed on the surface of the primitive earth.

Complex organic molecules containing up to eleven atoms, for example, ethyl alcohol ($CH_3 - CH_2 - OH$), are found in the interstellar dust clouds, and in comets and meteorites. Probably these molecules were synthesized during condensation of the planets or in solar nebulae even before the earth was formed. The contribution of these molecules to the initiation of life on earth seems unlikely.

10.4. PHYSICO-CHEMICAL LAWS CREATED NUCLEIC ACIDS AND PROTEINS

Although millions of tons of proteins are daily synthesized and degraded in our ecosystem by living species, the laboratory synthesis of proteins under prebiotic conditions and in aqueous solutions has only been partially successful. S. W. Fox united a mixture of 18 amino acids on a piece of volcanic lava and maintained the rock several hours in an oven at 170°C, thus creating the prebiotic conditions that prevailed on earth four billion years ago. By this procedure Fox synthesized various chains of polypeptides that are designated as *protenoïds*. Some of the protenoïds show a slight catalytic activity; according to Fox, the best catalysts were selected and, following a long evolutionary process, gave rise to enzymes, or biological catalysts.

The various amino acids could react together and form long chains of polypeptides only if the appropriate reactive sites of the two molecules were in intimate contact. Some minerals or clays could have facilitated these contacts by absorbing amino acids on their surfaces. Lahav demonstrated that such a situation occurred in the polymerization of the amino acid glycine in the presence of kaolinite. He carried out a series of evaporation-rehydration cycles with temperatures as high as 94°C, thus simulating a hot desert environment. Short polymers containing five monomers linked together by *peptide bonds* were produced in these experiments. Thus we may infer that, in principle, proteins may be synthesized from a mixture of amino acids in abiotic conditions.

Ponnamperuma and Sagan combined adenine, ribose, and phosphoric acid, a nonorganic substance, and irradiated the mixture with ultraviolet radiation, simulating solar energy. The reaction produced the all-important ATP. Nucleotides related to the three other nucleic bases were later synthesized by Ponnamperuma. It remained to demonstrate that nucleic acids could be synthesized by polymerization of the monomers, in the absence of today's complex cell machinery or any other catalysts. Schramm could obtain polynucleotide chains of less than 200 units by baking the monomers at 60°C in the presence of derivatives of metaphosphoric acid. Oro and Kimball used clay and cyanamide as a condensing agent and obtained sequences of 5 to 10 nucleotides.

Orgel and his coworkers showed that preexisting polyuridylates in an aqueous solution and in the presence of complementary monomers may serve as a template and direct the synthesis of new oligoadenylates according to Watson-Crick pairing rules. For example, short polymers containing 5 to 10 monomers of adenylates A are synthesized in the presence of a long polymer of complementary nucleotide polyuridylates.

Orgel's experiments are crucial, since they demonstrate experimentally the formation of an autocatalytic cycle of molecular "reproduction" in a purely chemical medium (see Fig. 10.5). This cycle implies that molecules include instructions for their own formation. The autocatalytic power of polynucleotides must be contrasted with the properties of polypeptides. The latter are not able to instruct their own formation; however, they are endowed with strong catalytic activity. Therefore, it is not surprising that the chemical symbiosis between polynucleotides in need of catalyst and polypeptides desperate for the conservation of "the blueprint of self" occurred, since both molecules could be formed contiguously in the same prebiotic environment.

The history of the first living cell is a long evolutionary process that originated from the mutual assistance of polynucleotides and polypeptides in their struggle for survival and reproduction. All the other organic molecules that were synthesized in the prebiotic soup, but were not able (because of purely chemical factors) to enter into such a symbiotic structure, eventually were degraded.

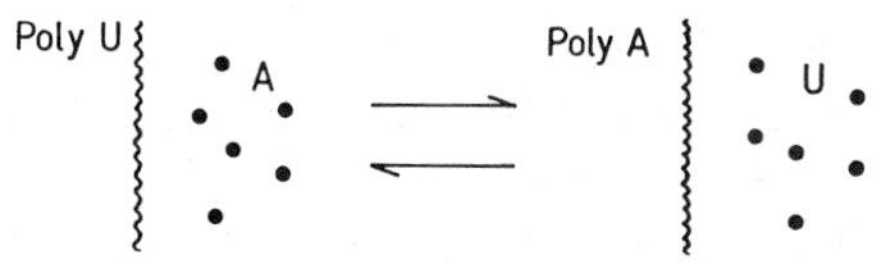

Fig. 10.5. Poly U is a template for the synthesis of Poly A which in turn directs the synthesis of Poly U.

10.5. DISSIPATIVE STRUCTURES AND BIOPOLYMER POPULATION ENHANCEMENT

Self-organized systems that are the result of nonequilibrium constraints and appropriate nonlinear couplings probably entered very early into the processes of the emergence of life from inanimate matter. Before any type of molecular evolution could be initiated, it was necessary that sufficient amounts of polymer become available in the primitive soup. The most common mechanism for polymer formation is the successive addition of monomer units in a chain elongation process. However, the free energy of the binding of monomers in the polymeric chain is such that the yield of the reaction is usually extremely low. On the other hand, polymers in water are hydrolyzed rapidly. Thus, once a polymer is formed, it is necessary that some mechanism enhance its population before complete annihilation. As we have already seen, synthesis on a template and the ensuing autocatalytic cycle may enhance polymer concentration substantially, provided the system operates far from thermodynamic equilibrium.

A. Goldbeter, G. Nicolis, and A. Babloyantz constructed several model polymerization processes which show the role of nonequilibrium and autocatalysis in the enhancement of polymer concentration.

Babloyantz considered a CSTR reactor open to the flow of two monomers, U and A, designating, respectively, the mononucleotide uracil and its complementary base adenine. mU and nA monomers may, in an extremely slow chain elongation process, eventually produce poly $U(B)$ and poly $A(P)$ strands. Adenine as well as all purine monomers in aqueous solution and at moderate concentration form n unit stacks designated by V^*. Very slowly, molecules in the stack may polymerize and give rise to a polyadenine P according to the following scheme

$$nA \rightleftharpoons V \tag{1}$$

$$V \rightleftharpoons V^* \tag{2}$$

$$V \rightleftharpoons P \tag{3}$$

$$mU \rightleftharpoons B \tag{4}$$

$$\tag{10.1a}$$

As soon as B, V^*, and P are formed, in addition to the chain propagation mechanism, an autocatalytic cycle is switched on, P becomes a template for the synthesis of B from mU monomers, and B acts as template for the condensation of stacks V^* which yield P templates. Finally, B and P may disappear by degradation, diffusion, or by forming stable Watson-Crick double-strand inactive molecules.

$$P + mU \rightleftharpoons P + B \tag{5}$$

$$B + V^* \overset{k}{\rightleftharpoons} P + B \tag{6}$$

$$B + P \rightleftharpoons BP \tag{7}$$

$$B \xrightarrow{\text{diffusion}} \tag{8}$$

$$P \xrightarrow{\text{diffusion}} \tag{9}$$

$$\tag{10.1b}$$

Experimental observations suggest that in this synthesis, once the first monomers are positioned on the surface of a polynucleotide template, the later addition of monomers becomes easier. Such a mutual assistance of monomers is called *cooperative binding*, and is also seen in biological catalysts. One may show that the cooperative property may be expressed by assuming that the kinetic constant of reaction (2) is a nonlinear function of V^*: $k = L\lambda^{V^*}$. λ and L are constants which specify the strength of a given cooperative behavior.

Assuming that the monomer concentrations are the constraints of the problem, the variables of the system are V, V^*, P, and B; therefore, the time change of concentrations again yields a set of nonlinear first-order differential

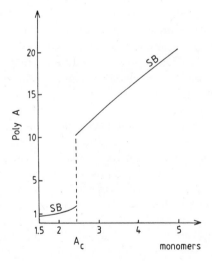

Fig. 10.6. Nonequilibrium constraints and template property enhance polymer production. At chemical equilibrium, the polymer yield is low. As monomer concentration increases beyond a critical value A_c, the fluctuations are amplified abruptly and the system switches to a new steady state showing a high polymer yield and chemical hysteresis. *SB* denotes a stable branch.

equations that, once solved, provide the amount of polynucleotide that is formed at each instant. Let us assume for simplicity that the reaction operates at a steady-state regime. Presently, we are interested in the change in poly A concentration as the monomer concentration increases.

The mathematical treatment of the problem follows exactly the same procedure as the one seen in Section 9.5, and shall not be repeated again. The outcome of such an experiment is shown in Fig. 10.6. One sees that for small monomer concentration, in the neighborhood of the equilibrium state, the autocatalytic cycle does not enhance the polymer yield. However, for appropriate rate constants and a critical value of monomer concentration, one can achieve a multiple steady polymerization state corresponding to the possibility of synthesis of small or large amounts of polymer for the same value of monomer concentration. Thus we see that only for a well-defined range of constraints, and for $A > A_c$, does the autocatalytic cycle exhibit its full potential. Beyond A_c the polymer concentration is increased dramatically and may accumulate in the medium, thus providing enough material for further evolution and complexification of the prebiotic system.

10.6. PREHISTORY OF THE FIRST LIVING ENTITIES: MOLECULAR EVOLUTION

In 1971 M. Eigen elaborated a coherent theory of prebiotic molecular evolution, which in later years was enriched with mathematical rigor and more importantly has been substantiated by experimental data. In essence, in a first stage Eigen extends the Darwinian ideas of evolution and selection to a population of macromolecules which appeared in the primitive soup; at a second stage Darwin's concepts are supplemented with the idea of cooperation between macromolecules. Finally, Eigen argues that a further evolution could not have been possible if the prebiotic systems did not compartmentalize as individual cellular units.

Eigen assumes that Darwinian-type macromolecular competition and selection was initiated in the midst of a large population of structurally distinct RNAs. These single-stranded molecules, in today's cell machinery, are double-duty agents with both functional and structural properties. The double-stranded DNA is only an instructional molecule which requires RNA for its own replication, and thus very likely is a latecomer in the prebiotic environment.

Not every RNA had a biologically meaningful and "correct" structure. To become a biopolymer, RNA must be able to fold itself such that stretches of complementary strands could form hydrogen bonds, thus stabilizing the molecular structure against the hydrolyzing power of water molecules. Such

RNAs were capable of self-replication, since they were templates which instructed their own synthesis. Therefore, according to the results of Section 10.5, their concentration could have been enhanced considerably, whereas the rate of synthesis of the "wrong" molecules could not overcome their rate of degradation.

Eigen's theory is based on a dynamic description of autocatalytic reactions involving the macromolecules of the prebiotic soup. The conclusions he draws are a direct consequence of the nonequilibrium nature of the autocatalytic processes.

10.6.1. Molecular Selection

Eigen's theory describes the events from the moment when a sufficient number of self-replicating RNAs with a template property were produced, following a scheme similar to Eq. (10.1), and had accumulated in the primitive soup. This population of self-replicating RNAs, all endowed with a template property, competed for the available "food," or energy-rich monomers.

The process of self-replication is not always perfect. Errors may appear in the positioning of one or several mononucleotides. When such errors are incorporated into the instructional macromolecules, we speak of the *mutation* of the original RNA. Mutations are currently seen in some of today's simple microorganisms.

Eigen developed a dynamic theory of these self-replicating molecules which will be summarized in the next two parts of this section. He showed that whenever mutations generate advantageous properties, a Darwinian-type evolutionary process operating at a macromolecular level selects the best competitor and eliminates the less fit molecular species from the environment.

Let us consider an RNA sequence i of N_i nucleotides. Each nucleotide position is identified by a subscript p that can take any value from 1 to N_i. If q_{ip} is the probability that the nucleotide at position p in sequence i is copied correctly during self-replication, then $(1 - q_{ip})$ is the error rate for that position. The probability that an entirely correct sequence i is replicated is the product of all the single nucleotide copying fidelities and is given by

$$Q_i = q_{i1} q_{i2} \cdots q_{iN_i} \tag{10.2}$$

Thus Q_i describes the quality of the template, that is, its ability to reproduce itself faithfully. The *quality factor* $Q_i \neq 1$ implies that replication of i is subject to errors. Because of errors, from i a new sequence j may be replicated with a finite probability.

In time the concentration X_i of species i may be enhanced for two reasons. If A_i is the rate of error-free self-replication of i from its own template, and if the probability of this event is Q_i, then X_i changes in time proportional to $A_i Q_i X_i$. Moreover, errors in replication of closely related molecular species may also give rise to the template i. Thus, a given replication process at template k produces either k at a rate $A_k Q_k X_k$ or molecules $j \neq k$ at a rate $A_k (1 - Q_k) X_k$; therefore,

$$\sum_i A_i (1 - Q_i) X_i = \sum_{i \neq j} \Phi_{ij} X_j \tag{10.3}$$

Concentration of X_i may also decrease in time with a rate D_i because of hydrolysis or a dilution flux factor proportional to Φ_{0i}. Thus the rate equations for X_i are

$$\frac{dX_i}{dt} = (A_i Q_i - D_i) X_i + \sum_{\substack{j=1 \\ (j \neq i)}}^{n} \Phi_{ij} X_j - \Phi_{0i} X_i, \qquad i = 1, \dots, n \tag{10.4}$$

If $E = A_i - D_i$ is defined as the rate of *excess production* of i, then $\bar{E} = \sum E_i X_i / \sum_i X_i$ is the average rate of excess production of all sequences present. And if $W_{ii} = A_i Q_i - D_i$, defined as the *selective value of* i, is the rate of correct replication of the sequence i, then one shows easily that Eq. (10.4) reduces to

$$\frac{dX_i}{dt} = (W_{ii} - \bar{E}) X_i + \sum_{\substack{j=1 \\ j \neq i}}^{n} \Phi_{ij} X_i, \qquad i = 1, \dots, n \tag{10.5}$$

with condition $\sum_{i=1}^{n} X_i = $ constant.

This system of coupled nonlinear differential equations may be studied subject to two types of constraints. Under the constant organization constraint, one assumes that the sum total of all species remains unchanged, whereas the constant flux constraint requires a uniform and continuous flux of monomers inside the system. The same general conclusions are drawn for both types of constraints, and therefore we shall only consider constant, overall organization.

From the first term of Eq. (10.5) we deduce that if for a given strand $W_{ii} > \bar{E}_i$, then X_i grows in time. However, if $W_{ii} < \bar{E}$, X_i decreases until sequence i disappears completely or is only produced by mutations represented by the second term of Eq. (10.5). The vanishing of any X_i increases the value of the average excess production $\bar{E}$. The surviving sequences now must overcome a larger value of $\bar{E}$. This property stems from the definition of $\bar{E}$. As $\sum X_i = $ constant, it is seen immediately that if one X_i disappears, the concentration of the remaining species with larger values of W_{ii} increases. Few templates fail these new requirements and disappear, thus increasing the value of $\bar{E}$ still further. Consequently, the number of competing species decreases, gradually; however, it does not reduce to a single strand, as no self-replication process is error-free and mutations constantly generate closely related molecules.

In the steady state that is eventually reached, a master copy wins the competition for "survival of the fittest." It must coexist with all mutant sequences derived from it by erroneous copying. Eigen argues that in the prebiotic soup the number of mutants surrounding a given "quasi-species" was extremely large, since the chemical kinetics of most mutant sequences could not have been much different from the kinetics of the master sequence itself.

A further investigation of the model described by Eq. (10.5) reveals that there is a threshold condition for the stable self-replication of a genetic message. If this threshold is exceeded, the genetic message cannot be conserved. The model predicts that the maximum gene length or the number of monomers in the RNA ranges from 50 to 100 nucleotides, which is similar to the size of today's transfer RNA.

This part of Eigen's theory transposes the Darwinian theory of the evolution of species to the realm of the macromolecules of the primitive soup. Only the species which could self-replicate themselves via a template mechanism, and thus had found a way to store and dispense information about their own structures, had some evolutionary advantage. For a given set of environmental conditions, only one quasi-species, which was best adapted, could emerge as a final winner. The selection of quasi-species was a purely chemical affair and was dictated by the ordinary laws of physics and chemistry. In order to survive the environmental conditions, the quasi-species became associated with other types of molecules which served as catalysts and helped the self-replication process. From a chemical point of view the polypeptides were such helpful auxiliaries and could have evolved simultaneously with the quasi-species.

10.6.2. Hypercycles

We have seen that the error threshold sets a maximum polynucleotide length limit of about 100 monomers. Such a short sequence cannot store a large repertoire of information. Nature must find a way to maximize Q_i, so that gene lengths of several thousand molecules could be formed. If the information content of genes is translated into proteins, or is associated with an existing protein endowed with a catalytic activity, the quality factor Q_i may be enhanced.

Let us consider a set of different master sequences together with their mutant distributions such that every member of the set taken alone is a winner team. If the members of the set could cooperate, instead of competing, a large quantity of information could be stored in the macromolecular system. Such cooperation may be achieved if quasi-species are connected together by mutual catalytic activity. Eigen calls such entities *hypercycles*. Figure 10.7a shows a hypercycle constructed from the cooperation of a set of two units. The hypercycle is formed from a self-replicating RNA (plus) and its complementary strand (minus) polynucleotides. The sequence I_1, codes for an enzyme E_1 which by virtue of its catalytic power helps the self-reproduction of RNA sequence I_2. This sequence in turn codes for a catalyst E_2 which helps the self-reproduction of RNA, I_1. Without such coupling only the "fittest" RNA would survive. The crosscatalytic coupling ensures the simultaneous survival and forced cooperation of both quasi-species, which function as a single unit with a larger informational content.

Using the theoretical methods of the preceding sections, Eigen and his coworkers investigated the dynamic coupling among quasi-species. They

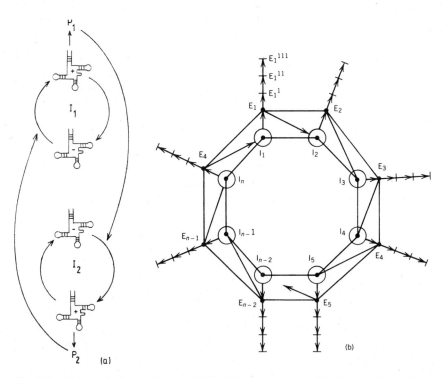

Fig. 10.7. (*a*) Instead of competing, two RNA sequences cooperate in a hypercyclic coupling. Information carrying RNA, I_1, codes for a primitive catalyst (enzyme) E_1, which in turn helps the self-replication of RNA, I_2. Such mutual cooperation stabilizes the information-carrying unit formed by $I_1 + I_2$ (Adapted from M. Eigen with permission). (*b*) An *n*-membered self-reproduction hypercycle. (From M. Eigen *et al.*, 1971, reprinted by permission Natur Wissenshaften.)

found that in self-replicating, closed-loop configurations and under certain circumstances, steady-state solutions in quasi-species concentrations may appear. This fact implies the coexistence of all quasi-species, and thus the formation of hypercyclic cooperation among the RNAs.

Hypercycles also compete with each other, but, contrary to Darwinian type self-replicating systems, selection leads to the survival of only one hypercycle. This "once for all selection" is final. Even if a new and more efficient hypercycle appears in small quantities, it cannot compete with a previously selected hypercycle. However, hypercycles may grow by incorporation of new RNAs and new proteins. Their structures may also evolve when new advantageous couplings arise in the hypercycles. Figure 10.7(b) shows an n-membered hypercycle.

10.6.3. Compartmentalization

In quasi-species populations, selection is based on purely chemical criteria, whereas in the winning hypercycle the message sent to the neighbors becomes important for self-enhancement and survival. However, this message does not evaluate the intrinsic fitness of the hypercycle in its environment.

So far in our scenario, all evolutionary processes have taken place in homogeneous media. Let us now assume that for some reason the prebiotic soup was distributed among separate compartments. In subdivided systems mutations could give rise to hypercycles of various evolutionary capabilities. In each compartment the evolutionary process may have switched back again to its Darwinian-type model. Compartments that were fitter in given environmental conditions could have been selected on the basis of their total performance. More and better genetic information provided a given hypercycle with a better chance of survival and made its further evolution possible. A given sequence of genetic information was improved by mutations and was transmitted to the next generation by the division of a compartment into two units, one holding the old strand while the daughter compartment enclosed the newly synthesized strand of polynucleotide (see Fig. 10.8).

Hypercycles and compartments dealt with two independent aspects of prebiotic evolution. Hypercycles insured the stable coexistence of self-replicating RNAs. Compartments provided a way for evaluation and improvement of information content; therefore, compartments dealt with the genotype-phenotype dichotomy.

Eigen continues his scenario of prebiotic evolution by assuming that the early genetic language was also based on a three-letter codon. He argues against the double-letter codon language by saying that nature wrote the book of life in the same language from beginning to end. It was neither economical nor practical to change the language in the course of biological evolution.

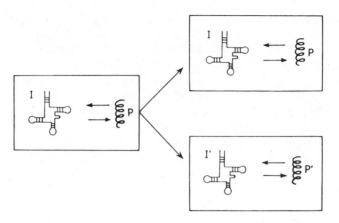

Fig. 10.8. In a compartment, information carrier RNA, I, is associated with its gene product P. During replication a mutant I' associated with a product P' is produced. Now if the system may be recompartmentalized into I/P and I'/P' daughter compartments, such a process allows selection of the RNA encoding the more efficient product. (After M. Eigen *et al.*, 1981).

The most fascinating aspect of Eigen's theory is that some conclusions drawn from it could be tested in laboratories today. The experience makes use of S. Spigelman's *minivariant* RNA that has been obtained by exposing an actual infectious virus to "ideal" conditions. In such a medium the normal virus sheds away about 4250 nucleotides and turns into a template of 220 nucleotides still capable of self-replication. These short, self-replicating RNAs are a convenient material for various evolutionary experiments that could be designed in order to test Eigen's theories. This story cannot be summarized here, and must be read in its original version.

In Eigen's theory it is assumed that evolutionary pressure somehow forced the homogeneous soup containing hypercycles into individual compartments. Such a compartmentalization process must necessarily arise under the action of physico-chemical laws of nature. The concept of spatial dissipative structures of Section 9.3.2 shows that it is possible to create, in a limited region of space, concentrations that may be much higher than in the homogeneous mixture. Moreover, the conditions for the spontaneous onset of dissipative structures are precisely those invoked by Eigen in the hypercycle model.

J. Hiernaux and A. Babloyantz associated catalytic polypeptides with the polymerization scheme of Eq. (10.1). They were able to show that, if the monomer concentration was inhomogeneous in the medium, mutants with the highest rate of replication, and therefore endowed with an evolutionary advantage, accumulated in the regions where the monomer concentration is the highest, whereas the slow-replicating molecules remained in the low monomer areas.

*   *

*

The material of the present chapter gives us some clues as to how a primitive cell could have emerged from inert matter. This self-organizational property of molecules was based on three important concepts: autocatalysis, non-equilibrium constraints, and the ensuing cooperative behavior of matter. In the next chapter we show the presence of dissipative structures inside present-day living cells.

Chapter 11

SELF-ORGANIZATION AT THE CELLULAR LEVEL

11.1. INTRODUCTION

Several of the self-organizational properties of the BZ reaction are seen in living cells. Rhythmic oscillations in concentrations and all-or-none-type phenomena are a few examples of such behavior.

Rhythmic phenomena in living systems may be seen at all levels of organization, with periods ranging from seconds to years. For example, as we shall see, concentrations of substances may oscillate in yeast cells with a period of the order of a few seconds. The most intimate and accessible short-period biorhythms are the regular seventy heartbeats per minute. This complex activity may be traced ultimately to molecular events.

Practically all *eukaryotes* (multicellular organisms) from vegetals to animals have a built-in (*endogenous*) rhythmicity of 20 to 28 hours. In the absence of all external influence, these circadian rhythms (*circa* = environment, *dies* = day) have a stable periodicity. The underlying mechanism of these oscillations is certainly again of a biochemical nature. Even a unicellular alga, *Acetabularia mediterranea*, shows a circadian rhythm in its oxygen balance. As we have seen in the BZ reaction, oscillating phenomena were generated by regulatory interactions between different molecules. Biochemical clocks are governed by similar laws, and therefore we expect that the same mathematical approach may describe a large variety of rhythms in biological organisms.

Two examples of dissipative structures inside cells are given in this chapter: the glycolytic oscillations in yeast cells and enzyme induction in bacteria. In the case of glycolytic oscillations, the periodicities are generated by the activity of biocatalysts called *enzymes*.

11.2. THE CONCEPT OF THE CELL

In 1839 Schwann introduced the concept of the "cell," defined as the unit of biological organization. The cell, he said, is a bag full of chemical substances that interact according to the ordinary laws of physics and chemistry. In his time urea and acetic acid, two essentially biological substances, had already been synthesized in laboratories. Therefore, vitalistic theories had to give way to *reductionist* biochemistry.

With today's powerful electron microscopes our view of the cell has evolved considerably. A cell is not just a bag of molecules, but is a highly complex and structured entity. Figure 11.1 shows an animal cell from a multicellular organism.

The *nucleus* is the command center of the cell. All information and instructions concerning the vital biochemical processes and cell reproduction are stored in the nuclei, imprinted in the genetic material and supported by DNA molecules. *Ribosomes* are the protein factory of the cell. They function according to the instructions received from nuclei via a nucleic acid called *messenger RNA. Mitochondria* are the respiratory agents of the cell and perform all oxidative chemical reactions of the unit. They are power plants

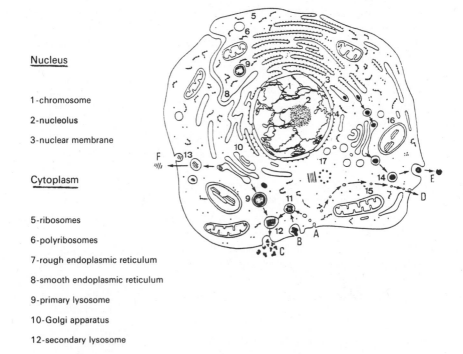

Nucleus

1-chromosome

2-nucleolus

3-nuclear membrane

Cytoplasm

5-ribosomes

6-polyribosomes

7-rough endoplasmic reticulum

8-smooth endoplasmic reticulum

9-primary lysosome

10-Golgi apparatus

12-secondary lysosome

15-mitocondrion

17-centrioles

Fig. 11.1. A eukaryote cell. The cell membrane encloses a complex and organized structure made of a nucleus and various organelles. (From P. Van Gansen, *Abrégé de Biologie Générale* © Masson 1984.)

that regenerate the energy-rich molecules from less energetic ones. *Golgi apparatus* packages and stores the glycoproteins, while *lysosomes* are preformed packages of enzymes which catalyze various digestive reactions.

All these organelles are surrounded by membranes and are constituted from smaller subunits which have their own internal organization. Finally, these organelles are embedded in the *cytoplasm*, a gel-like structure which is composed mainly of proteins. The cytoplasm itself is a complex structure of microtubules and microfilaments; only small molecules, such as O_2, CO_2, or sugars, diffuse easily in the cytoplasmic gel.

In each cell thousands of vital chemical processes take place, according to a well-defined space and time sequence. Cells import food, which is "burned" and yields energy or is degraded into biomaterials which are used for the synthesis of new bioproducts. These processes involve many chemical steps and biological catalysts. For example, the meat (protein) we eat is first degraded into its constituent amino acids before new proteins are synthesized from them.

One designates by *metabolism* the sum total of the various chemical reactions occurring in a cell which produce energy or synthesize new biologically important macromolecules. The chain of reactions which transforms a given substance into a final product is called the *metabolic pathway* of that substance. The intermediate products of a metabolic pathway are called *metabolites*.

In the present chapter we are interested in the metabolic pathway of glucose, the ordinary sugar molecule. This pathway consists of a long chain of chemical reactions. In each step a metabolite is transformed by a specific *enzyme*. At the same time, energy is released, or is stored in chemical bonds of a specific molecule, ATP, which, as we saw, is related to the constituents of the genetic material. Thus our first task is to investigate the nature and the mode of functioning of enzymes. We must also understand how energy is stored in an ATP molecule.

11.3. ENZYMES OR BIOLOGICAL CATALYSTS

Nearly all biochemical reactions are performed in the presence of specific catalysts called *enzymes*. These catalysts enhance the rate of reactions a millionfold, as one enzyme may catalyse 10^6 molecular events in one minute. Evolution had to create enzymes; without them, the temperature inside living organisms is too low for any chemical transformation to occur at a significant rate.

Ferments were the first known biological catalysts, and in 1878 the name *enzyme* (leaven) was given to all of them. Fifty years later it was realized that enzymes were proteins made of a long chain of amino acids linked by peptide bonds (see Fig. 10.2).

Usually a given enzyme catalyzes one particular biochemical reaction by acting only on one specific reactant. The latter is called the *substrate* of the enzyme. Substrates bind only to specific regions of the enzyme, called *active sites*, and form *enzyme-substrate complexes* (*ES*). An active site is a three-dimensional entity formed by the coiling of the amino acid chain. These cleft or crevice-type structures take up only a small part of the total volume of the enzyme.

The specificity of an active site is determined by the precise arrangement of atoms forming the cleft. In a *lock and key* model a substrate enters its corresponding cleft in the same way that a key fits into a lock. However, some active sites are not rigid, the shape of the site being modified by the binding of the substrate. In such an *induced-fit* model, only after the binding of the substrate does the active site take a shape complementary to the substrate (see Fig. 11.2).

In the experimental study of biochemical reactions, one measures the re-action velocity for different values of substrate concentration—while the total amount of enzyme is held constant. Figure 11.3 shows that a reaction velocity increases linearly with substrate concentration. However, at high substrate values a saturation effect is seen which appears only in enzyme-catalyzed reactions. In this region every enzyme molecule is busy making or breaking chemical bonds, and therefore the velocity may not increase, regardless of how much more substrate is added to the reaction medium.

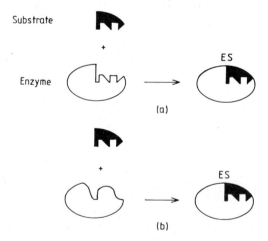

Fig. 11.2. (*a*) Lock and key model of enzyme substrate interaction. (*b*) In the induced-fit model, only after substrate binding does the active site of the enzyme take the proper complementary shape.

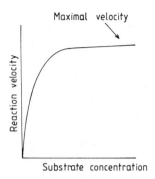

Fig. 11.3. Kinetics of an enzyme-catalyzed reaction. Velocity versus substrate concentration.

The activity of enzymes is regulated by several mechanisms. Some enzymes are synthesized as *inactive precursors* which later are transformed into active catalysts whenever their presence is required in the cellular metabolism. Catalytic activity of an enzyme may be reduced or annihilated completely by specific small molecules called *inhibitors*. Beneficial drugs and toxic agents act by inhibiting specific enzymes.

Some enzymes are inhibited by the product of the reaction they catalyse. Reaction (11.1) shows a substrate S which is transformed by successive enzymes $E_1, \ldots, E_n$ into a final product P. The latter may inhibit the activity of E_1 or any of the intermediary enzymes; in this case we speak of *feedback inhibition*.

$$S_1 \xrightarrow{\;E_1\;} S_2 \xrightarrow{\;E_2\;} S_3 \cdots S_n \xrightarrow{\;E_n\;} P \qquad (11.1)$$

A substrate may bind to the active site of the enzyme by electrostatic interactions, hydrogen bonds, or other attractive forces.

Allosteric Enzymes. The important class of allosteric enzymes comprises several identical subunits or *protomers*, each endowed with its own active site. In these enzymes the binding of the substrate to one of the subunits may influence positively the catalytic activity of the sites in other subunits; thus the binding of the substrate may become highly cooperative.

Kinetics of allosteric enzymes often do not follow the graph of Fig. 11.3, as may be seen from the plot of reaction velocity versus substrate concentration shown in Fig. 11.4. The sigmoïdal curve contrasts with the usual saturation type of kinetics. An allosteric enzyme is recognized from such S-shaped cooperative kinetics found from experimental data. Allosteric enzymes have

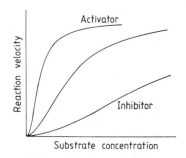

Fig. 11.4. Kinetics of an allosteric enzyme. Activators of the enzyme enhance, whereas the inhibitors decrease, the reaction velocity.

been described by two distinct theoretical models. We start with the concerted model of Monod, Wyman, and Changeux.

Allosteric enzymes are formed from several subunits. For the sake of simplicity let us consider only two identical subunits, each having one active site. Suppose that in the absence of any substrate the enzyme may exist in either of two conformations, designated by R (relaxed) and T (tense). The T state has a low affinity for the substrate, whereas the R state binds the substrate much more readily. (see Fig. 11.5). Moreover, the two conformations are in equilibrium, and nearly all free enzyme molecules are in the T form. The important assumption of the model is that the two subunits of the same enzyme must necessarily adopt the same conformation. The RR and TT states are allowed, whereas the RT conformation is forbidden. As the substrate only binds to R state, the addition of substrate shifts the conformational equilibrium in the direction of the R form. When the substrate binds to one site, the second site of the enzyme must also switch to the active state R. Thus the T-R transition is *concerted*. As more substrate molecules are added, the number of enzyme molecules in the R state increases nonlinearly in a *cooperative* fashion.

Allosteric binding of substrate to the enzyme is described by the following scheme

$$R_0 \rightleftharpoons T_0$$
$$R_0 + S \rightleftharpoons R_1$$
$$R_1 + S \rightleftharpoons R_2 \tag{11.2}$$

Again, by our usual methods, one finds easily that

$$V = V_{max} \frac{S}{K_R} \left[\frac{1 + S/K_R}{L + (1 + S/K_R)^2} \right] \tag{11.3}$$

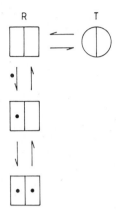

Fig. 11.5. The reversible transition between T and R conformation is fully concerted. The binding of the first substrate ($\bullet$) switches the conformation to the high affinity form RR.

V_{max} is the maximum velocity of the enzyme. This expression is a highly nonlinear function of substrate concentration S and accounts for the characteristic sigmoïdal kinetics of allosteric enzymes. In this expression $L = T_0/R_0$ is the ratio of substrate-free enzyme concentrations, whereas $K_R = 2R_0 S/R_1$ is the dissociation constant of the enzyme-substrate complex. The ratio L is also called the *allosteric constant*, and its value characterizes the degree of cooperativity of the enzyme.

Some inhibitors use this cooperative behavior in order to deactivate the allosteric enzymes more efficiently. The *activators* of these enzymes may also enhance their catalytic power by a similar mechanism.

Allosteric activators bind preferentially to the R conformation, whereas inhibitors show a high affinity for the T form. The binding of activators or inhibitors shifts the $R \rightleftharpoons T$ equilibrium in the desired direction. The effect is expressed quantitatively by a change in the value of L. Inhibitors decrease L whereas activators increase its value (see Fig. 11.4). Sigmoïdal kinetics of allosteric enzymes may also be accounted for by the *sequential model* of D. Koshland. The descriptions of the model may be found in appropriate biochemistry books.

11.4. THE BIOLOGICAL ENERGY STORAGE UNIT: ATP

In some respects living organisms are essentially complex machines that cannot function without constant energy input from the environment. Plants obtain the required energy from the sun and store it as free energy in the

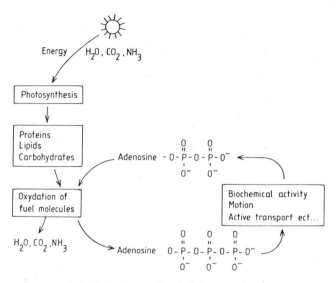

Fig. 11.6.   Energy flow in living organisms.

chemical bonds of ATP molecules by a process known as *photosynthesis* (see Fig. 11.6). In the absence of photosynthesis, living organisms must extract energy from foodstuffs. Here again the free energy is stored in ATP molecules made of an adenine, attached to a ribose sugar and a triphosphate unit. The free energy is stored in the oxygen-phosphate bonds of the molecule. In the presence of water, the phosphate units are hydrolyzed and adenosine diphosphate, ADP, or adenosine monophosphate, AMP, is formed. The free energy of the phosphate bond is liberated and is used to drive biochemical processes which require energy input or to perform mechanical work, such as muscle contraction.

ATP is the "gold standard" in the economics of the living kingdom. It is a universal and immediate free energy donor, and has a very high turnover. In a fast tennis game we may use 0.5 kg of ATP per minute, whereas at rest our consumption may be 40 kg in a 24-hour period.

ATP is constantly regenerated from ADP. The process requires photosynthesis or the burning of biofuel (see Fig. 11.6). The ordinary sugar glucose is one of the important biofuels that generate ATP. This is why nothing can match a sugar cube for a tired organism. We will now investigate in more detail the transformation of glucose in organisms and the production of ATP molecules. In other words, we are interested in the details of the *glycolytic pathway*.

11.5. THE GLYCOLYTIC PATHWAY

In the framework of self-organizing systems, our interest in the glycolytic pathway stems from the fact that the concentration of intermediates of glucose degradation in yeast cells exhibits oscillations reminiscent of the chemical clock behavior seen in the BZ reaction.

Glucose transforms into pyruvate according to the following stoichiometry and produces two energy-rich ATP molecules:

$$\text{glucose} + 2P_i + 2ADP + 2NAD \longrightarrow$$
$$2(\text{pyruvate}) + 2ATP + 2NADH \tag{11.4}$$

NAD is the nicotinamide-adenine-dinucleotide molecule, and NADH is the same molecule binding a hydrogen atom. This global reaction encloses many intermediate steps, each catalyzed by a specific enzyme, as shown in Fig. 11.7.

According to environmental conditions, pyruvate molecules may follow three different paths. In the absence of oxygen, for example in muscle cells, pyruvate is transformed into *lactate*. In mitochondrial membranes, pyruvate molecules, after some transformation, combine with oxygen and yield CO_2 and water. This pathway is known as the *respiratory chain*. Cells obtain twenty times more energy from the respiratory chain than from glucose-pyruvate transformation. (Let us note that if glucose is burned outside the organism, it also finally is transformed into CO_2 and H_2O.) In the third pathway some microorganisms, such as yeast cells, transform pyruvate into ethanol, the ordinary drinking alcohol. In what follows, our interest will be focused on the glycolytic pathway of yeast cells.

Yeast cells are unicellular microorganisms, some of which live on grape skins and are the agents of transformation of grape juice into wine. The process of fermentation produces alcohol and also some 500 different organic components. The number and composition of these products transform the grape juice into a superb Mouton Rothschild or a cheap wine. Yeast cells can transform sugars of any origin into ethanol, and thus it is not surprising that alcoholic beverages were known to all societies from the dawn of civilization. The first written account of glycolysis in the form of a narrative series of drawings was found in the grave of the keeper of the wine cellar of Thoutmosis III, an Egyptian Pharao who lived from 1505 to 1450 B.C.

Wine often turned into vinegar, causing economic disaster. Therefore, research was initiated in order to understand the mechanism of fermentation. However, only in 1837 was it realized that the process of fermentation was not spontaneous but was mediated by living yeast cells. The study of these fermentation reactions enabled Louis Pasteur to show that life stemmed only

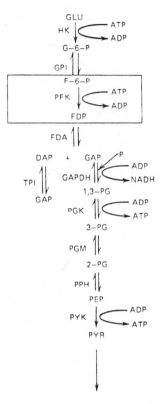

Fig. 11.7. The glycolytic pathway. Framed reactions represent the part of the scheme which has been modeled.

from life. He also isolated components other than ethanol from the wine.

In 1897 Buchner realized that cell extracts were as efficient fermenting agents as the intact cells. This discovery proved for the first time that fundamental biological processes may proceed outside the living organism. Later work of Embden and Meyerhof showed that glycolysis is not restricted to microorganisms, but is a general mechanism for the utilization of glucose by other cells.

The elucidation of the glycolytic pathway opened the doors of the living organism to chemistry, and the new science of *biochemistry* was born. Life had lost a part of its mystery and uniqueness. By 1940 the entire pathway as we know it today was elucidated. In 1964 another discovery surprised the scientific community: under specific experimental conditions, all the intermediates of the glycolytic pathway showed regular oscillations. As we have shown in Part II, this discovery coincided with the development of the theory of self-organized systems. Several theoretical models for glycolytic oscillations were elaborated by E. E. Selkov, J. Higgins, and A. Goldbeter and R. Lefever. Since the allosteric model of Goldbeter and Lefever (1972) seems to fit the experimental data most closely, it will be reproduced here. But let us first describe the oscillatory phenomenon in more detail.

11.6. THE GLYCOLYTIC PATHWAY IS A BIOLOGICAL CLOCK

Under suitable conditions, regular concentration oscillations of glycolytic intermediates are observed in a single or synchronized population of yeast cells. Cell-free extracts prepared from yeast or muscle cells metabolize glucose following exactly the same metabolic pathway as did the intact cells. Cell extracts are very convenient tools for experimentation. In their presence biological processes of life are reduced to ordinary chemical reactions, and can therefore be studied more easily. A reconstituted glycolytic system has been studied by B. Hess and A. Boiteux (1971).

In a typical experiment, glucose is injected at a constant rate into a reactor containing cell extracts, thus creating a controllable open chemical system. Oscillations are most conveniently observed by recording the fluorescence of one of the intermediates of the pathway. Figure 11.8 shows such a measure-

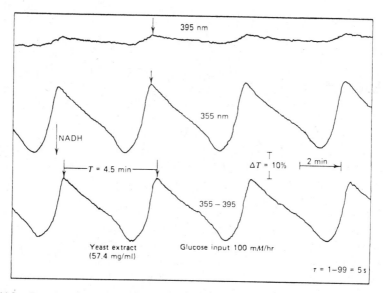

Fig. 11.8. Experimental measurements of one of the intermediates of the glycolytic pathway (NADH), showing periodic oscillations. (From Hess and Boiteux, 1971. Reproduced, with permission, from the *Annual Review of Biochemistry*, Volume 40, © 1971 by Annual Reviews, Inc.)

ment. Modification in the fluorescence intensity of NADH indicates a concentration change of this metabolite. Oscillations are observed only in a narrow and low range of substrate input rates. Outside this range all intermediates are produced at a steady-state regime.

All glycolytic intermediates oscillate with the same period but with different phases. They may be classified in two groups separated by a constant phase difference. The period of oscillation is of the order of minutes and is temperature-dependent. The concentration of metabolites oscillates between 10^{-5} and 10^{-3} moles per liter. Moreover the rate of substrate entry determines the shape and amplitude of oscillations. The periodic phenomena may be of quasi-sinusoidal or non-sinusoidal form, or exhibit double periodicity.

11.6.1. The Molecular Basis of Oscillatory Behavior

A good scientist must have the perspicacity and sharpness of a first-rate detective. Faced with a large number of suspects, an experimentalist must cross-examine all available data before deciding "Who did it." Such a procedure has been applied to the numerous metabolites and enzymes of the glycolytic pathway. It is found that the fructose-6-phosphate (F6P), which is the substrate of the enzyme phosphofructokinase (PFK), is the last component of the chain that, if acted upon, produces oscillations. Metabolites downstream with respect to PFK do not influence the oscillatory behavior of the pathway. Other experimental data point to the fact that PFK is the key regulatory enzyme which generates the oscillatory behavior of the entire glycolytic pathway. Moreover, the reaction catalyzed by PFK is virtually irreversible. As a rule, in metabolic pathways these types of reactions are potential sites of control.

PFK is found to be an allosteric enzyme made from several identical subunits. Each subunit may bind simultaneously, but at distinct sites, F6P and ATP molecules, which are transformed into ADP and FDP (fructose 1, 6 diphosphate). In the presence of low concentrations of ATP the reaction velocity as a function of F6P exhibits ordinary enzyme kinetics (see Fig. 11.3). However, at high levels, ATP is an allosteric inhibitor of PFK and the reaction kinetics show the characteristic sigmoidal curve of Fig. 11.4. The inhibitory action of ATP is reversed by direct activation of the enzyme by its product ADP. The periodic "on-off" performance of the catalyst is the result of the allosteric nature of the enzyme. In the glycolytic pathway (Fig. 11.7), PFK is inhibited by the substrate ATP but is activated by the product ADP. This fact brings about a regular change in enzyme activity, which in turn is reflected as an oscillatory behavior in the metabolite concentrations. The catalytic power of PFK is found to vary between 1% and 80% of its maximum activity.

In what follows we will construct a model glycolytic pathway and show that the oscillations in metabolite concentrations arise as a result of the instability of the homogeneous steady state of the system. These dissipative structures are the biological counterpart of the BZ reaction and function as chemical clocks.

11.6.2. An Allosteric Model for Glycolytic Oscillations

The construction of a model for the glycolytic pathway would seem to be a gigantic task. One must deal with many coupled nonlinear differential equations and a great number of parameters. However, several experiments suggest that oscillations start at the level of the reaction catalyzed by the enzyme phosphofructokinase. This result and the findings of the preceding subsection indicate that it might be possible to model the oscillatory character of the glycolytic pathway by isolating and considering only one reaction from the entire metabolic chain (Fig. 11.7). The following reaction seems to be the best candidate:

$$F6P + ATP \xrightarrow{PFK} ADP + FDP \qquad (11.5)$$

Let us remember that the key fact about this reaction is that PFK is an allosteric enzyme made of several identical subunits. Each unit has its own distinct binding sites for products and reactants. Therefore, it is assumed that oscillations are produced as a result of the regulatory action of the enzyme activity. In a given range of glucose concentrations, the enzyme is under the influence of two antagonistic factors. PFK is inhibited cooperatively by ATP and is activated by ADP. Moreover, experience shows that the concentration ratio of F6P/FDP does not influence the oscillatory behavior, whereas the ratio ATP/ADP is a determining factor. Therefore, we may simplify the model still further and assume that only one substrate (ATP) is transformed by the enzyme into a single product (ADP):

$$ATP \xrightarrow{PFK} ADP \qquad (11.6)$$

The problem may be simplified further, without any prejudice to the global property of the model, if the multi-subunit protein is replaced by a "model" dimer enzyme formed only from two subunits.

Even with all these simplifications, because of the allosteric nature of the enzyme, reaction (11.6) is still very complex, as it represents globally a great number of elementary steps. The detailed scheme is shown in Fig. 11.9. Enzymes with a different number of substrates or products, bound to the active sites, are considered as separate intermediate species.

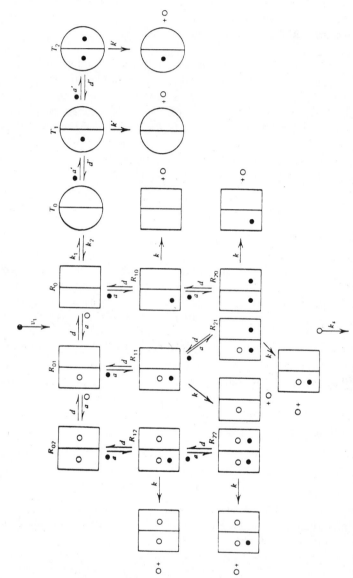

Fig. 11.9. Concerted model for a dimer enzyme activated by the reaction product. Substrate (•) enters the reactor at a constant rate v_1. Product: (○) binds only to the high-affinity form R and is removed from the system proportionally to its concentration. a_i, d_i are association and dissociation constants of the enzyme complex. k, k', are reversible decomposition rate constants of R and T form.

   Time-dependent variables of the problem are substrate concentration S, product concentration P, and nine R_{ij}s representing enzyme species in the R state that bind i molecules of S and j molecules of P (i and j take values of 0, 1, and 2). T_0, T_1, and T_2 are the concentrations of free or enzyme substrate complex in the T state. Therefore, fourteen nonlinear coupled ordinary differential equations are necessary for the dynamic description of the reactions depicted by the scheme shown in Fig. 11.9. Moreover, as the glycolytic oscillations are also seen in homogeneous media, the diffusion of metabolites is not considered.

   The enzyme is a catalyst, it is not consumed by reaction processes, and therefore the total sum of all enzymatic forms is a constant quantity D.

$$\sum_{i,j=0}^{2} R_{ij} + \sum_{i=0}^{2} T_i = D = \text{const.} \tag{11.7}$$

Moreover, the total quantity of enzyme is much smaller than the amount of substrate it must transform into product. Consequently, the enzymatic species reach a quasi steady state very rapidly; therefore all dR_{ij}/dt and dT_i/dt vanish. These conditions together with the conservation relation, Eq. (11.7), reduce the twelve differential equations to a set of coupled nonlinear algebraic equations which may be solved easily and yield enzyme values as a function of substrate and product concentrations. These values are substituted in the remaining two differential equations for substrate and product concentrations. Thus, the investigation of glycolytic oscillations reduces to a discussion of two coupled first-order nonlinear differential equations.

   These differential equations may take a much simpler form if expressed in *reduced* variables. These are obtained by dividing the actual variables, the concentrations of ATP and ADP, by appropriate constants of the problem. If α designates the reduced substrate and γ the reduced product variables, differential equations of the problem take the following form:

$$\frac{d\alpha}{dt} = \sigma_1 - f(\alpha, \gamma)$$

$$\frac{d\gamma}{dt} = f(\alpha, \gamma) - k_s \gamma \tag{11.8}$$

Here σ_1 is proportional to v_1 and is a measure of constant substrate input and k_s is the rate constant of product elimination from the reaction medium.

   All cooperative kinetics of the allosteric enzyme appear in the function $f(\alpha, \gamma)$:

$$f(\alpha,\gamma) = \frac{\sigma_M \left[(1+\gamma)^2 \left(\dfrac{\alpha}{\varepsilon+1}\right)\left(1+\dfrac{\alpha}{\varepsilon+1}\right) + \theta L\left(\dfrac{\alpha c}{\varepsilon'+1}\right)\left(1+\dfrac{\alpha c}{\varepsilon'+1}\right) \right]}{L\left(1+\dfrac{\alpha c}{\varepsilon'+1}\right)^2 + (1+\gamma)^2\left(1+\dfrac{\alpha}{\varepsilon+1}\right)^2}$$

In this function α and γ are coupled nonlinearly and are modulated by the allosteric constant L. c, ε', and ε are related to the various kinetic constants of the global reaction. The constant $\theta = k'/k$ is the ratio of the turnover numbers of the T and R states. σ_M is proportional to the maximum rate of the enzyme reaction.

The two coupled differential equations (11.8) are members of the same family of mathematical objects as Eq. (9.4) which describe the chemical clock embodied in the Brusselator scheme. Therefore, from this point on we follow the same procedure as the one developed in Part II.

First we evaluate the time-independent concentrations α^s (ATP) and γ^s (ADP) at steady state as functions of different parameters of the problem. These functions may be single-valued or exhibit multiple steady states. We test the linear stability of these states by introducing small perturbations $\delta\alpha$ or $\delta\gamma$ in the system. Evolution of these perturbations are followed in time; if they increase in an oscillatory fashion then the full nonlinear system may also exhibit oscillatory solutions. This may happen only if the steady state is an unstable focus.

The detailed analysis of the allosteric model described by Eqs. (11.8) shows that, for a well-defined range of substrate injection rates, the steady state is an unstable focus. The direct integration of these equations shows regular and periodic change of PFK activity in time (see Fig. 11.10a). The plot of ATP versus ADP is a limit cycle which surrounds the unstable focus.

This limit cycle is shown in Fig. 11.10b and indicates that both concentrations of ATP and ADP oscillate in time. If the value of the various parameters of the system are chosen in accordance with the existing experimental data, the substrate and product concentrations oscillate between a high value of 10^{-3} and low value of 10^{-5} mole per liter. The period of oscillations is of the order of one minute. Both the amplitude and the period of oscillations are influenced by a change in the values of substrate input rate σ_1, total enzyme concentration D, and the rate of product elimination from the reacting medium k_s.

All these results are in qualitative and quasi-quantitative agreement with the known experimental data of glycolytic oscillations. It is worthwhile to note that the agreement between theory and experience was obtained by modelling a complex, many-variable pathway by a simple, two-metabolite scheme. Moreover, a vital pathway for energy storage in cells was studied as a simple

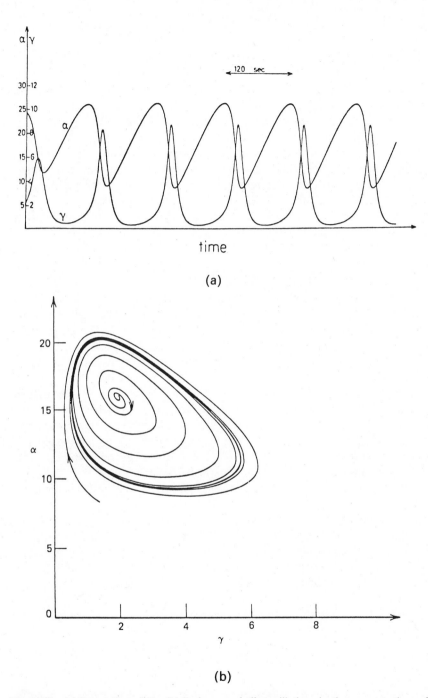

Fig. 11.10. (*a*) Integration of Eq. (11.8) shows periodic oscillations in the concentrations of γ (ADP) and α (ATP). (*b*) Trajectories in the phase plane α-γ (ATP-ADP) evolve toward a limit cycle.

open chemical system, kept far from thermodynamic equilibrium by the action of PFK. The cooperative nature of the allosteric enzyme PFK gave a nonlinear character to the dynamical rate equations which described the functioning of the metabolic pathway.

We have already seen on several occasions that openness, irreversibility, and nonlinearity were the essential characteristics of dissipative structures. Only under these conditions may a homogenous steady state become unstable and bifurcate into a self-organizing state. Thus we may conclude that, subject to a well-defined range of substrate injection rates, the glycolytic pathway is a dissipative structure that self-organizes itself into a biological clock.

In the next section we consider another type of dissipative structure which is of prime importance in the metabolism of *Escherichia coli*.

11.7. MEMORY EFFECTS IN BACTERIA

Escherichia coli is one of the commonest and best known bacteria. These harmless unicellular creatures live in the human gut. They reproduce by dividing every twenty minutes or so. Thus a single *E. coli* may produce thousands of progeny in three hours time and a billion bacteria in a day.

As yet it is not clear why this bacterium has chosen our hospitality, though it has been a fortunate event for science. These small, fast-growing protein factories used in large quantities are an excellent protein supplement in husbandry. But above all *E. coli* has been a perfect experimental material for the investigation of the secrets of life. These tiny bacteria are one of the most comprehensively studied creatures on earth. Most of our major concepts of biochemistry and genetics, such as the modulation of protein synthesis, and the decoding of genetic language, emerged from the study of *E. coli*, and especially from a well-defined portion of its DNA called the *lac operon*.

The aim of the present section is to show that the vital biosynthetic processes that are initiated at the level of the lac operon may be understood in the framework of dissipative structures. In the manner of glycolytic oscillations, we reduce the complex biochemical events to a simple, open chemical system which exhibits cooperative behavior. We show that an appreciable amount of a given protein is synthesized only when the system bifurcates into a more active state.

But first let us sketch in broad lines the mechanism of protein synthesis from amino acid building blocks.

11.7.1. The Machinery of Protein Synthesis

The deciphering of the mechanism of protein synthesis took a lot of ingenuity from many researchers. However one must single out the important contributions of J. Monod and F. Jacob for the elaboration of the lac operon model of *E. coli*.

We have seen that the genetic message coding for the synthesis of proteins is securely wrapped in the double helicoïdal structure of the DNA molecule. However, DNA is not the direct template for protein synthesis. Protein molecules are assembled by the ribosomes (see Fig. 11.11), which are small protein factories embedded in the cytoplasm and usually located at some distance from DNA. (In multicellular organisms, DNA is enclosed in the cell nuclei.)

The genetic information flows from DNA into the cytoplasmic pool of amino acids by the intermediary of ribonucleic acids. Three distinct type of RNA molecules participate in protein synthesis. The blueprint of a given protein, engraved in its corresponding gene, is *transcribed* from DNA into an RNA which has a base sequence compatible with the gene sequence of the DNA. This single-stranded *messenger RNA* travels toward the ribosomes. These little factories are made of proteins and RNA molecules arranged in small subunits. Ribosomes are nonspecialized structures which can synthesize any protein that has been ordered by DNA via its messenger mRNA.

The various amino acids required by mRNA are brought to the ribosomal assembly line by specific *transfer* RNAs. These cloverleaf-like molecules are endowed with an amino-acid attachment site and a three-letter *anticodon* region that is recognized by mRNA. The anticodon forms complementary base pairs with the codon of the amino acid which must enter into the protein chain. Thus tRNAs speak simultaneously DNA and amino acid language and are translators in the synthetic processes of the cell.

For a short time each tRNA in turn, trailed by its corresponding amino acid, attaches itself to the ribosomes. Meanwhile, the ribosomes travel along the

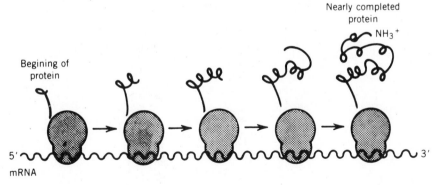

Fig. 11.11. A protein factory. Ribosomes move along the mRNA. They function independently of each other. (From *Biochemistry*, 2nd ed. by L. Streyer, W. H. Freeman and Company. Copyright © 1981).

mRNA and read the message which gives the order of attachment of various amino acids. The latter are assembled into proteins by appropriate enzymes. In short, protein synthesis requires the coordinated interplay of more than a hundred macromolecules, a good part of which are activating enzymes. In summary, protein synthesis requires the flow of the genetic information according to the following scheme:

$$\text{DNA} \xrightarrow{\text{transcription}} \text{RNA} \xrightarrow{\text{translation}} \text{protein} \qquad (11.9)$$

11.7.2. *Escherichia coli* Copes with Environmental Constraints

E. coli can use lactose, a form of sugar, as its unique source of energy. With the help of enzyme *β-galactosidase* it splits lactose into glucose and galactose (see Fig. 11.2). The various enzymes of the glycolytic pathway transform the glucose into energy-rich molecules and necessary metabolites. Each bacterum needs several thousand copies of β-galactosidase in order to live on lactose. However, if bacteria are grown directly on glucose as a sole source of food, the number of β-galactosidase per cell is fewer than ten. Thus β-galactosidase is an

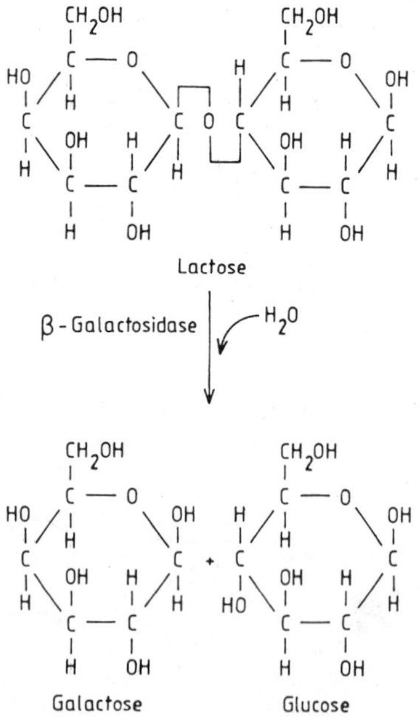

Fig. 11.12. Enzyme β-galactosidase hydrolyzes lactose into a galactose and a glucose unit. Permease cannot discriminate between lactose and IPTG, which is a gratuitous inducer.

inducible enzyme; it is only synthesized by *E. coli* in the presence of lactose molecules.

At a first stage, the few existing β-galactosidase molecules that are present prior to induction modify the lactose into a related sugar, *allolactose*. This new molecule is the actual inducer that acts upon the genetic material and brings about the synthesis of new enzyme molecules.

Much to the delight of researchers, some sugars may induce β-galactosidase, but are not metabolized by the enzyme. In the presence of these *gratuitous* inducers, the process of induction may be isolated from the complex metabolic pathways of the cell and studied separately.

11.7.3. The Molecular Basis of Enzyme Induction

The DNA of *E. coli*, enclosing all necessary information for cell survival and reproduction, is made of a closed chain of nucleotides. The contour length of the DNA is about a thousand times as long as the greatest diameter of the bacterium. Only a small portion of the DNA codes for the induction of β-galactosidase. This stretch of the genome is made of three adjacent elements, as shown in Fig. 11.13.

The regulatory gene *i* instructs the synthesis of a tetrameric protein called a *repressor*. In the absence of an inducer, this protein interacts with the *operator gene o*. This part of the DNA controls simultaneously the three neighboring *structural genes z, y*, and *a*. The control site *o* and its associated structural genes form the *lac operon*. The binding of the repressor to the operator gene blocks simultaneously the transcription of the three structural genes. The *promoter site p* binds the enzyme RNA polymerase. The inducers, natural or

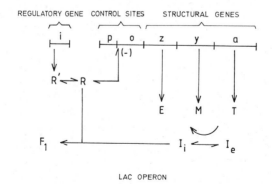

Fig. 11.13. Enzyme induction. The regulatory gene codes for a repressor R. The structural genes code for three proteins: β-galactosidase (E), permease (M), and transacetylase (T). Repressor R binds to the operator and prevents the transcription. Inducers bind to the repressor and keep the operator free.

gratuitous, bind to the repressor, thus preventing it from blocking the operator gene. The three structural genes z, y, and a are transcribed into a single mRNA molecule which in turn codes simultaneously for all three proteins. The z gene codes for galactosidase, y for a protein called *permease*, and a instructs the synthesis of *thiogalactosidase transacetylase*. Permease is a vehicle for the transport of lactose across the bacterial cell membrane, whereas the function of transacetylase in cell metabolism has not yet been elucidated.

The salient feature of the induction process is the feedback generated by the action of permease. This transport protein facilitates the entry of the inducer into the cell; the latter increases the concentration of permease (see Fig. 11.13).

11.7.4. The Induction Process Has a Memory

The most important elements of β-galactosidase induction, after the addition of a gratuitous inducer, have been incorporated by M. Sanglier and A. Babloyantz in the following simple model:

$$R' \underset{k_1'}{\overset{k_1}{\rightleftarrows}} R$$

$$R + O^+ \underset{k_2'}{\overset{k_2}{\rightleftarrows}} O^-$$

$$R + n_I I_i \underset{k_3'}{\overset{k_3}{\rightleftarrows}} F_1$$

$$\eta + O^+ \overset{k_4}{\rightarrow} O^+ + E + M$$

$$M + I_e \underset{k_5'}{\overset{k_5}{\rightleftarrows}} M + I_i$$

$$M \overset{k_6}{\rightarrow} F_2$$

$$E \overset{k_7}{\rightarrow} F_3 \tag{11.10}$$

The regulatory gene i produces an inactive repressor R'. The latter is in equilibrium with an active repressor R which is a four-unit allosteric protein. R binds to the operator O^+ and blocks its expression by forming the complex O^-. If n molecules of inducer I_i are present, the repressor is deactivated cooperatively and a substance F_1 is formed. Once the operator O^+ has been liberated, the structural gene may function and two proteins E and M are synthesized from a pool of amino acid denoted globally by η. Permease M helps the entry of external inducer I_e into the cell, and E represents β-galactosidase. The last two steps account for decay or dilution of enzyme and permease. The model must be supplemented by the condition that, within the cell, the total amount of operator $O^- + O^+ = X$ is conserved.

The scheme of Eq. (11.10) is a minimal model that accounts for the qualitative behavior of the induction process. In a more realistic description, the third and fourth steps must be replaced by an allosteric model similar to the one used for glycolytic oscillations. Moreover, the transport of inducer by permease is a much more complex process than the one described by the fifth step. Nevertheless, simple as it may be, the scheme accounts in a satisfactory manner for the salient features of β-galactosidase induction.

Five nonlinear coupled differential equations, in variables $R, O^+, E, M,$ and I_i, are necessary for the description of the time evolution of the induction process. The theoretical approach we consider is identical to the one used for the study of glycolytic oscillations in Section 11.6. Therefore, we shall only discuss the final results of the model.

In an imaginary experiment we start with bacteria which have been grown solely on glucose, and thus are not induced. We transfer them into a medium containing only a gratuituous inducer called IPTG. In the presence of a fixed concentration of IPTG, if we wait long enough the induction process becomes time-independent. A steady state is reached that may be evaluated by solving the time-independent evolution equations of the model. Thus we are led to the numerical solution of a set of ordinary coupled nonlinear algebraic equations. The solution of these equations furnishes the value of β-galactosidase

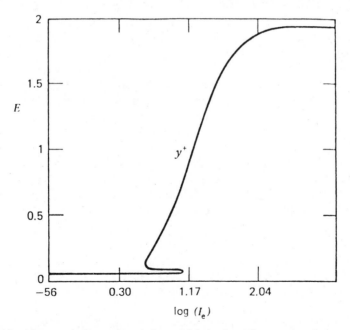

Fig. 11.14. All-or-none phenomenon in bacterial induction. The quantity of enzyme versus external inducer concentration shows multiple steady states. Bacteria operating on the lower branch are non-induced. In a given culture and in a given range of concentration, both types of bacteria may coexist.

concentration in terms of the various parameters of the system. More specifically, the concentration of the enzyme appears as a nonlinear function of the concentration of the extracellular inducer I_e.

Fortunately, some rate constants and few parameters entering the scheme of Eq. (11.10) may be evaluated from experimental data or guessed by indirect methods. With the help of these parameters, the change in the steady-state concentration of induced enzyme E, in response to the extracellular concentration of inducer I_e, has been calculated numerically (see Fig. 11.14). In accordance with experimental data, the model predicts that for small inducer concentrations the level of E is very low. However, for a threshold value of I_e, multiple steady states appear. Only three states are possible in a narrow range of inducer concentrations. As the latter increases, a single steady state with a high level of enzyme concentration remains. The linear stability analysis predicts that the upper and lower branches are stable, while the middle branch is unstable. Thus for a critical value of external inducer, bacteria may either remain in a weak state of induction, characterized by the lower branch of Fig. 11.14, or be fully induced with enzyme concentration, as shown by the upper branch.

Numerical values also compare favorably with the experimental results. From Fig. 11.14 one finds that in non-induced bacteria $E = 2 \times 10^{-3}\,\mu M$; this figure must be compared with the experimental value of $3.3 \times 10^{-3}\,\mu M$; whereas, in the induced state, the theoretical and experimental values of E are respectively $2\,\mu M$ and $3.3\,\mu M$.

With this simple model we have shown that protein synthesis induced by external factors in a bacteria is a dissipative structure of the type seen in Schlögl's model of Section 6.7. Whenever the inducer concentration exceeds a threshold value, the nonequilibrium constraints, the feedback, and nonlinear dynamics drive the system into a cooperative behavior.

In the presence of natural inducers, the induction process may exhibit oscillatory behavior. A theoretical model similar to Eq. (11.10) has been constructed by M. Sanglier and G. Nicolis. They show that, for given parameters, three states are available to the bacteria: a stable steady state, and an unstable limit cycle which is surrounded by a stable limit cycle.

In the glycolytic pathway, metabolic oscillations were generated by the regulation of existing enzyme activity, whereas in β-galactosidase induction the activity of the enzyme is regulated by the pulsatory synthesis of the catalyst. We thus speak of *epigenetic oscillations*. Metabolic and epigenetic oscillations are important in cell economy and cellular productivity. Both phenomena are dissipative structures which only arise because of the feedback mechanisms that bring about the instability of a homogeneous steady state.

In living organisms dissipative structures may be seen at every level of organization. In the next chapter we show their importance for an understanding of the aggregation of a colony of isolated cells.

Chapter 12

CELLULAR COMMUNICATION

12.1. MEMBRANE STRUCTURE AND INTERCELLULAR COMMUNICATION

The two examples of the preceding chapter were mainly concerned with events inside a unicellular organism. However, in the process of bacterial induction, the cells did receive inducers from the outside world, and the transport of the latter across the membrane was partly responsible for the all-or-none coherent molecular behavior.

In complex multicellular organisms, coherence is the rule. Many different cellular types must cooperate in order to form organisms or give rise to regular heartbeats or generate circadian rhythms. The cooperation of millions of brain cells which initiates thought, feelings, and action is certainly the ultimate in biological coherence.

In the present chapter multicellular cooperation is illustrated by an example showing how unicellular amoebae communicate with each other, signaling food scarcity and community gathering in order to ensure reproduction. We show that such cellular communication results from the instability of the homogeneous solution of an appropriate dynamical system. But first let us examine more closely the molecular mechanism of intercellular communication.

12.2. CELL MEMBRANES

In order to achieve cooperation, cells must communicate—cellular messages must cross cell boundaries; therefore, the present chapter starts with a summary of our current view of cell membranes and various mechanisms of information transfer across these barriers.

Cells are surrounded by sheet-like structures called *membranes* of about 10^{-6} cm thickness. These are highly selective permeability barriers which separate "inside" from "outside" and give cells their individuality. In eukaryotic cells, organelles, such as mitochondria, or nuclei are also surrounded by membranes.

Biological membranes are the traffic controllers and telecommunication agencies of cells. They exercise a tight control over the nature and the amount of the influx and outflux of molecules. They receive and transmit messages inside the cell. Membranes are also involved in vital biosynthetic processes. The green color of plants is due to chlorophyll molecules which are stored in tiny organelles inside cells, called *chloroplasts*. Membranes of chloroplasts capture the energy of the sunlight and transform it into free energy stored in the chemical bonds of *carbohydrates* that are formed by photosynthetic processes from CO_2 and water. All free energy used by organisms may ultimately be traced to these carbohydrates. The fuel molecules of cells are also oxidized by mitochondrial membranes to yield ATP.

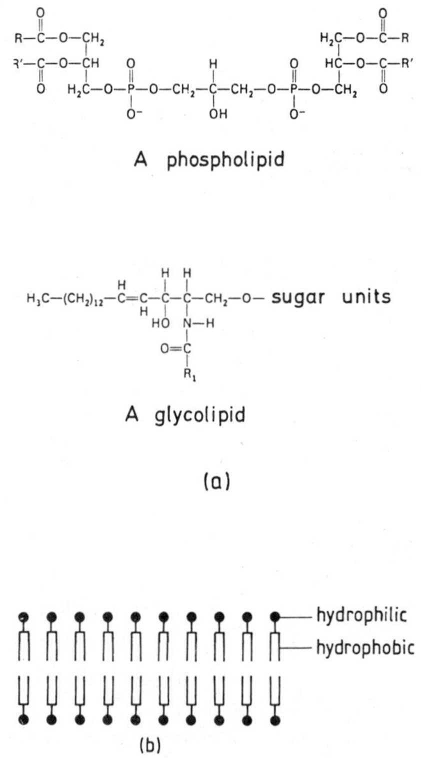

A   phospholipid

A   glycolipid

(a)

— hydrophilic
— hydrophobic

(b)

Fig. 12.1. (*a*) Two common membrane lipids. (*b*) Phospholipids and glycolipids in aqueous media form bimolecular sheets. The hydrophilic polar heads face the aqueous medium, while the hydrophobic tails are sequestered inside the bilayer.

12.2.1. Membrane Structure

Membranes are formed mainly from proteins and lipids (fats). Lipids are a very important class of biomolecules. They are highly soluble in organic solvents and insoluble in water. Lipids form the membrane canvas and are also highly concentrated energy reservoirs. The unpopular cholesterol is an example of a lipid molecule. The most common membrane lipids are glycolipids and phospholipids (see Fig. 12.1a). They both contain a hydrophylic (water-loving) part attached to the hydrophobic (water-repelling) moiety. For example, in phospholipids the large hydrocarbon tail is insoluble in water, whereas the electrically charged head is water soluble. A minute's reflection shows, and experience confirms, that a good way to separate two aqueous media is to form a lipid bilayer, as shown in Fig. 12.1b. This is exactly what nature did in its attempt to compartmentalize biochemical events in the primitive soup.

However, cell survival necessitates exchange of substances with the environment, so a natural biological membrane cannot be totally imperme-

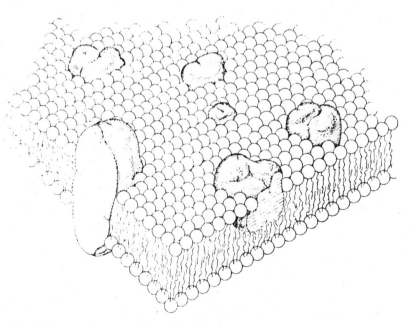

Fig. 12.2. Fluid mosaic model. In biological membranes proteins are embedded in the lipid bilayer and ensure distinctive membrane permeability and function. (From S. J. Singer and G. L. Nicolson. *Science*, Vol. 175, February 13, 1972, pp. 720–723. Copyright © 1972 by the AAAS.)

able. Thus various types of protein molecules are embedded in the lipid bilayer in order to insure distinctive membrane permeability and function (see Fig. 12.2). The nature of these proteins specifies a given membrane. Embedded transport proteins serve as *gates*, *channels*, and *pumps*. Other proteins form receptors used by cells for capturing molecular messages from the outside world. Some membrane proteins are enzymes that help photosynthesis or ATP formation. The membrane structure is fluid; lipids and proteins diffuse in the plane of the bilayer. Membranes are structurally asymmetric: the two surfaces of the bilayer are not identical, and they are also different functionally.

12.2.2. Pumps

Some of the transport proteins of the membrane may function in the same fashion as a mechanical pump. In animal cells these pumps maintain a high intracellular concentration of K^+ and a low concentration of Na^+. The flow of these ions proceeds against their concentration gradient; ions must climb uphill. Therefore, an expenditure of energy is required in this *active transport* process, and is furnished by ATP molecules. Each protein pump extrudes three sodium ions, and simultaneously two K^+ ions enter the cell. Therefore this transport system is called $Na^+ - K^+$ *pump.*

A model based on the dual conformation of the pump proteins has been proposed to explain the functioning of the transport system. One protein conformation binds Na^+ preferentially and the binding site of this ion faces inside the cell. The ATP molecules furnish the required energy and trigger a conformational change to a state which binds K^+ preferentially. The binding

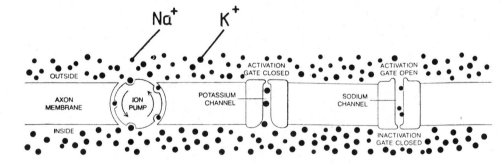

Fig. 12.3. Transport across a nerve cell membrane. The membrane separates outside fluid, which is about 10 times richer in sodium ions (small dots) than potassium ions (large dots), from inside fluid, which has the reverse ratio. The membrane is penetrated by proteins that act as selective channels for sodium and potassium transport. In a resting cell, these channels are closed. An ion pump maintains the ionic imbalance by pumping out sodium ions in exchange for potassium ions. (From C. F. Stevens, 1979, reprinted by permission from *Scientific American*.)

site of K^+ faces outside the cell. The sequence of these events is depicted in Fig. 12.3. Usually one pump exchanges 200 Na^+ ions against 130 K^+ ions every second. A typical brain cell contains 1 million such pumps.

12.2.3. Channels

Other membrane proteins form channels that are selectively permeable to various molecules (see Fig. 12.3). Sodium channels are permeable to Na^+ and largely exclude K^+ ions. Potassium channels do the reverse. Channels open and close either in response to a difference of electrical potential (K^+ ions are electrically gated) or as they are influenced by chemical substances.

12.2.4. Receptor Activation

We have already seen that specific protein molecules are embedded in the lipid bilayers and act as information capturing centers. In this mode of communication, a mediator molecule, for example, a hormone, carries the message to the specific membrane proteins called *receptors* which in turn transfer the information and generate an intracellular response, such as protein synthesis or enzyme activation or inactivation (see Fig. 12.7).

Cyclic AMP is a common intracellular messenger molecule which is formed from energy-rich ATP. The enzyme *adenyl-cyclase* chops two phosphates from ATP and twists the third into a ring molecule, as shown in Fig. 12.7. Another enzyme, *phosphodiesterase*, destroys the message conveyed by cAMP. The enzyme transforms the latter into AMP by opening the newly formed ring.

Information mediators are usually small molecules containing from 10 to 100 atoms. Receptors, on the other hand, are gigantic aggregates of atoms, the three-dimensional assemblage of which determines receptor specificity.

Receptor and mediator interaction is highly specific. However, receptors may be fooled and made to bind foreign molecules that look chemically similar to the mediator. The harmful agents in some diseases and also beneficial drugs may act by the saturation of natural receptors. A dramatic example is furnished by cholera, which is mediated by *Vibrio cholerae*. The latter would be absolutely harmless if they did not secrete a chemical that acts as a mediator for specific receptors of the organism. Under normal conditions, for a proper digestion of nutrients by appropriate enzymes of the small intestine, an alkaline medium is necessary. In the presence of digestible material a mediator acts upon corresponding receptors and two liters of alkaline fluid flushes into the intestinal tract. At the end of the digestion period, the fluid is restituted to the organism by the colon. In cholera patients the intestinal receptors are constantly activated by the fake mediator, the bacterial toxin, and 20 to 30 liters of water enter the intestine. The organism is unable to recover such an amount of fluid, which is eliminated by diarrhea and vomiting. If not treated the patient dies from dehydration.

12.2.5. Gap Junctions

The interior of cells may also directly communicate via special molecular arrangements that span the intervening spaces, or gaps, between opposed cell membranes. These intercellular tunnels are called *gap junctions*. They provide rapid flow of inorganic ions and metabolites from cell to cell.

*   *

*

After this short survey of various modes of cellular communication, let us go back to the special case of communication between individual cells of a colony of amoebae, *Dictyostelium discoideum*, commonly known as *slime molds*.

12.3. SLIME MOLD AGGREGATION

Presently we shall treat an example in which receptor activation is the major means of intercellular communication. The communication process has been studied in some detail in colonies of the cellular slime molds, *D. discoideum*. The life cycle of the amoeba is illustrated in Fig. 12.4.

The cycle begins with spores (seeds) which germinate into unicellular amoebae whenever the environmental conditions become favorable for the species. In the presence of large quantities of bacteria, their favorite nutrient, they live an independent vegetative life and multiply by dividing into two daughter cells. When all bacteria are consumed and amoebae are starving, some individual *pacemaker* cells begin, after a few hours, to secrete periodic pulses of cAMP into the medium. Pulses of cAMP are emitted every few minutes and are signals for community gathering. The velocity of the chemical signaling wave is of a few microns per second. The amoebae are attracted by cAMP and move toward the signaling centers. We speak of *chemotaxis*, meaning movement in response to a chemical stimulus. The signal is relayed by incoming amoebae, thus extending the aggregation territory. Finally, up to a hundred thousand amoebae aggregate around a given pacemaker cell.

Aggregation shows time and space periodicity. Successive steps of inward cellular movement appear to propagate outward from the center toward the periphery of the aggregation territory (see Fig. 12.5). At the end of the aggregation process, a conical mass rises from the aggregate, continues to elongate and eventually topples over like a *slug*, and starts to move about. At

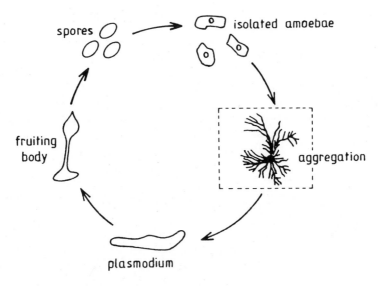

spores   isolated amoebae
fruiting body   aggregation
plasmodium

Fig. 12.4. Life cycle of amoebae. In favorable conditions spores develop into isolated amoebae. Bacteria-starved, amoebae aggregate and form a slug which finally rises into a fruiting body and produces spores.

the same time, wavelike contractions of a periodic nature are seen in the body of the slug. Finally, the slug stops moving and a stalk is formed, surmounted by a mass of spores. After formation of this fruiting body, a new cycle may begin.

In the fruiting body, spore and stalk cells are different functionally and chemically, although both types are formed from the same amoebae. We say that cells are *differentiated* in order to form an organism.

In following the sequence of events depicted in Fig. 12.4, we have witnessed a *developmental process* that culminated in *morphogenesis* (the birth of a form). With only two cell types, *D. discoideum* is one of the simplest possible multicellular organisms. Much work has been devoted to the elucidation of the different stages of the life cycle of these amoebae. However, the seemingly simple morphogenetic events involve a long series of complex chemical and morphological changes that has only been partially elucidated.

We shall leave the problem of morphogenesis for the next chapter. For the moment, let us focus on the developmental events which bring about cell aggregation around pacemakers. In other words, let us look more closely at the succession of events in that part of Fig. 12.4 which is delimited by a rectangle. Fortunately, experimental data are quite abundant in this area, and this has facilitated the construction of theoretical models.

12.3.1. The Experimental Facts

Our first concern is to investigate the molecular mechanism of cell response to cAMP message. Experiments are usually performed with cell suspensions or with bacteria grown on agar plates. Constant or pulsatory input of cAMP may be introduced locally into the experimental medium by a needle mimicking a pacemaker. Cell reactions to these signals are followed in time.

The onset of starvation marks the zero point of our chronology. From this moment on, in a population of isolated amoebae four major types of cell behavior are seen.

1. Shortly after the onset of starvation, cells secrete little or no cAMP. Moreover, at this stage amoebae are indifferent to the presence of external cAMP.

2. Later, the individual cells acquire the ability to relay the signal they receive. At this stage, cells respond to the external cAMP by synthesizing and secreting a cAMP pulse in order to attract other cells.

3. Some eight hours after the onset of starvation, a small number of privileged cells function as pacemakers and secrete periodically, every few minutes, a pulse of cAMP. The acquisition of pacemaking ability is an intrinsic property of the individual cells and in nature arises spontaneously in the absence of any external stimulation.

Fig. 12.5(a)

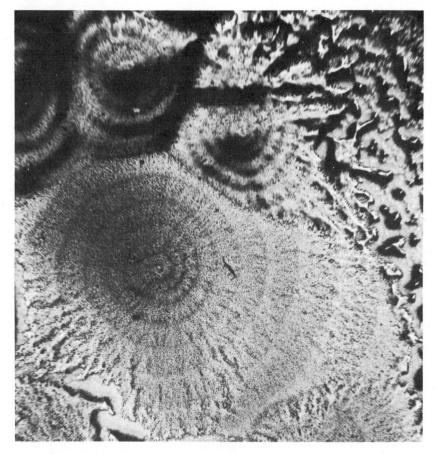

Fig. 12.5(b)

Fig. 12.5. During the aggregation process, wave patterns of chemotactic activity is seen in colonies of slime molds. (*a*) Note the spiral waves rotating in opposite direction. (From F. M. Ross *et al.*, 1981, reprinted by permission *J. Gen. Mic.* (*b*) Details of one spiral. Courtesy of G. Gerisch.)

4. The cells at the center of an aggregate secrete cAMP at a high and steady level.

The time sequence of these events is depicted in Fig. 12.6. Any good model must describe and explain, on a molecular basis, the behavior of amoebae from starvation time up to the pacemaking stage. In the example of glycolytic oscillations, we saw how a change in enzyme activity may bring about complex and diverse behavior in the system. Then the question arises whether, in the

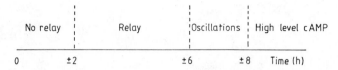

Fig. 12.6. Experimental observations show that in the hours after the onset of starvation, a given cell may exhibit four different behaviors.

present case, enzyme versatility is not again the key to the understanding of developmental processes. Goldbeter and Segel took this option and constructed a model that explains qualitatively the time behavior of slime mold aggregates. They have assumed that there is a unique molecular signaling system which can function sequentially in the four different ways depicted in Fig. 12.6.

However, in order to be able to construct a model, we need to know the nature of the chemical species and the mechanism of their interactions.

12.3.2. The Molecular Basis of the cAMP Signaling System

Experimental data indicate that the signaling system in *D. discoideum* amoebae comprises two proteins incorporated in the lipid bilayer of cell membranes. A receptor protein binds cAMP and is functionally coupled to the enzyme adenylate cyclase, as shown in Fig. 12.7. The receptor faces the exterior of the cell, while inside the cell the internally oriented enzyme adenylate cyclase transforms ATP into cAMP. The binding of external cAMP

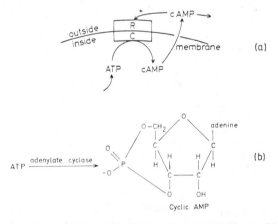

Fig. 12.7. (*a*) A two-unit protein is embedded in the cell membrane. The outside moiety R is the receptor, and the inside half C is the enzyme adenylate cyclase. (*b*) The enzyme transforms ATP into cAMP which is transported into the extracellular medium, which activates the receptor (+).

to its receptor increases the catalytic activity of the inside enzyme. Newly synthesized cAMP is secreted to the extracellular medium where it may activate more receptors and ultimately is hydrolyzed by the enzyme *phosphodiesterase*. Thus the cAMP synthetic process is self reinforcing, or, according to our definitions, autocatalytic.

The major assumption of the model is that the various modes of functioning of this signaling system bring about the four stages of the amoebae life cycle as depicted in Fig. 12.6. From Fig 12.7 we can see that concentrations of extracellular and intracellular cAMP are certainly "good" variables for any model. These are two independent variables, as the extracellular cAMP may be modified by external factors, while internal cAMP is entirely dependent on internal cellular machinery. ATP may be considered as another intracellular variable, since it is the substrate for cAMP synthesis.

Following our usual methods, we must write the laws which govern the change in time of the various variables in the problem. In other words, rate equations must be constructed for the three variables, extracellular cAMP, intracellular cAMP, and ATP. As the life cycle unfolds, the solutions of these equations, which are time-dependent functions, describe the evolution of the signaling system.

Rate equations become much simpler if we define new variables by dividing the concentrations of ATP, intracellular cAMP, and extracellular cAMP by various constants of the problem and call them *reduced variables*. We find

$$\frac{d\gamma}{dt} = (k_t\beta/h) - k\gamma$$

$$\frac{d\beta}{dt} = q\sigma\varphi - k_t\beta$$

$$\frac{d\alpha}{dt} = v - \sigma\varphi - k'\alpha \tag{12.1}$$

The reduced extracellular concentration of cAMP is denoted by γ and may change in time for two reasons. Cells secrete pulses of cAMP into the external medium. The cAMP molecules must somehow leave the membrane bilayer, and we have seen that these processes may be extremely complex. However, in the present problem we focus mainly on the signaling system; therefore, we assume simply that the more there is inside the more there will be outside. In this approximation $k_t\beta$ designates the amount of cAMP that enters the extracellular medium per unit time, and the constant h is a dilution factor. On the other hand, phosphodiesterase destroys cAMP at a rate of k molecules per unit time, and again, the more there is the more will be destroyed. This process introduces the $-k\gamma$ contribution in the rate equations.

The variable β designates the intracellular concentration of cAMP, and $-k_t\beta$ accounts for the transport of these molecules across the cell membrane. We have seen that intracellular cAMP synthesis from ATP is mediated by adenylate cyclase, which is functionally coupled to the receptor protein which binds extracellular cAMP. Thus we are in the presence of an autocatalytic process. The exact nature of the autocatalysis and the mode of action of the enzyme upon its receptor are not presently known. Therefore, some plausible hypothesis must be made at this level.

We assume that the nature of the phenomenon we are about to describe is very similar to the substrate-product interaction of an allosteric enzyme as seen in the glycolytic oscillations. Modified to suit the present situation, the allosteric model is assumed to function in the following manner. The receptor-enzyme complex is a membrane-bound protein made of four subunits. Two subunits, each endowed with a cAMP binding site, function as receptors, and the remaining two subunits act as adenylate cyclase. We assume a concerted allosteric interaction between the subunits which can exist in two conformational states. From this point on, the model parallels the procedure we followed for glycolytic oscillations. Again a quasi-steady state assumption is made for all enzymatic and receptor subunits. After some cumbersome calculations, one finds that the intracellular cAMP production is given by

$$\varphi = \alpha(1 + \alpha)(1 + \gamma)^2 / [L + (1 + \alpha)^2(1 + \gamma)^2] \qquad (12.2)$$

Here L is the allosteric constant. In Eq. (12.1), σ is the maximum activity of adenylate cyclase. The variable α designates the concentration of ATP, and q is some function of the pertinent kinetic constants. The functional form of φ reveals that the rate of production of intracellular cAMP is a highly nonlinear function of ATP and extracellular cAMP.

In the third part of Eq. (12.1) the concentration of α may change in time for three reasons. ATP is synthesized in the organism by various metabolic pathways. We shall not go into the details of these processes, but simply assume that a constant quantity of ATP flows per unit time to the site of the membrane where all the action is located. A constant v term in the rate equation of α describes this process. As we have seen, ATP is transformed into cAMP by a highly nonlinear process. Calculations show that, in order to account for this loss, a contribution proportional to $-\sigma\varphi$ must be added to the rate equation of α. ATP is also used in many other metabolic pathways of amoebae. We denote this fact by a loss term $-k'\alpha$ proportional to α and modulated by a global rate constant k'.

It is important to notice that all the nonlinearities in Eq. (12.1), and therefore their analytical properties, are the consequence of the functional form of φ. The latter is the direct result of the positive feedback exerted by extracellular cAMP and the hypothesis about the allosteric nature of the receptor-enzyme complex. Many other assumptions concerning the enzyme-receptor interaction may be made which furnish a completely different φ, however, without altering qualitatively the results.

Before we investigate the content of Eq. (12.1), let us go back to the known data about the activity of the two key enzymes of the signaling system. Let us remember that σ and k are, respectively, the activities of adenylate cyclase and phosphodiesterase, as shown in Fig. 12.8. As the life cycle of amoebae unfolds, the representative point of the system travels in the σ-k plane.

These facts suggest that we must integrate the differential equations of our problem for various values of σ and k. These parameters are, respectively, a measure of adenylate cyclase and phosphodiesterase activity. The other parameters of the system are fixed at some reasonable values deduced from plausible experimental considerations.

The set of Eq. (12.1) can be analyzed by our usual methods. Linear stability analysis of a three-variable system requires much more labor but still is feasible. Therefore, one is able to determine the values of different parameters which push the system into an unstable region. Once these values are found, the differential equations may be integrated numerically, and yield the desired

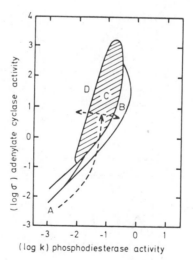

Fig. 12.8. The partitioned adenylate cyclase (σ) and phosphodiesterase (k) activity plane. Domains B, C, and D correspond, respectively, to relay capability, autonomous oscillations, and high steady secretion of cAMP. After starvation, the representative point of the system moves slowly along the developmental path (broken arrow). Upon ceasing to oscillate, the system may either return to the excitability domain B or enter the domain of high steady cAMP secretion. (From A. Goldbeter and L. Segel, 1977.)

values of $\alpha = \alpha(t)$, $\gamma = \gamma(t)$, and $\beta = \beta(t)$. The latter may be related immediately to the sequence of events of the life cycle of amoebae.

Computer integration shows that if we construct a graph of σ versus k as shown in Fig. 12.8, four separate regions appear with markedly distinct properties.

In region A, where σ and k are small, the solution of the differential equations shows that cells secrete cAMP into the external medium regularly and at a low level. As σ and k increase, region B is reached. In this state, if the steady-state solution of the differential equations is perturbed by the addition of a minute quantity of cAMP, a new solution appears, as seen in Fig. 12.9b. The cell amplifies the small perturbation and delivers it to the external medium in the form of a pulse and returns to the previous quiescent state. We recognize immediately a relay mechanism in this process. For a higher and appropriate ratio of σ and k, region C is reached. Here the perturbed steady state bifurcates into a new solution which exhibits sustained oscillations of the limit cycle type, as already seen in the glycolytic pathway (see Fig. 12.9c). For still higher values of σ we enter domain D. The solution of differential equations shows only a steady and high level of cAMP concentration. We thus see a one-to-one correspondence between the time sequence of Fig. 12.6 and the journey in parameter plane $\sigma - k$, as shown by the broken-line arrow.

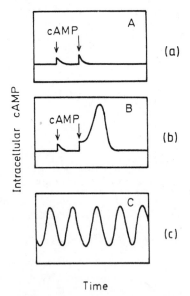

Fig. 12.9. Solution to the differential equations (12.1) with $\sigma - k$ parameters corresponding to the three domains of Fig. 12.8. (b) Propagation of a pulse of cAMP. (c) Autonomous oscillations. (From A. Goldbeter and L. Segel, 1977.)

The aggregation process may be summarized in the following manner. A population of amoebae in a typical laboratory experiment may contain some 10^5 independent cells. These individual cells may have slightly different metabolisms. Therefore, amoebae start and travel the developmental sequences of Fig. 12.6 at different moments and with different speeds. In region A amoebae are isolated and indifferent to any extracellular signal. Those who first reach region C individually become pacemakers. Moreover, only cells in region B are able to relay the signals they receive from the pacemakers. Finally, amoebae that reach region D become part of the slug tip which is rich in cAMP.

It is important to note that the model has kept its promises and explains one part of the developmental sequences of the life cycle of amoebae.

*   *

*

The aggregation mechanism prepares the amoebae for the differentiation process. In the slug they become either stalk or spore cells. Our next example shows how dissipative structures may account for the processes of differentiation. The theory we shall develop is general, and is not limited to slime mold aggregation.

Chapter 13

THE DEVELOPMENT OF ORGANISMS

13.1. INTRODUCTION

This chapter discusses the theoretical study of some aspects of *developmental biology*. In the present context, the concept of development describes the ensemble of events that transform a single fertilized egg or seed into a fully grown adult organism. The process is awesome. From a tiny cell, gigantic organisms, such as an elephant or a man, are born. Different tissues are formed, and their cooperation enables the organism to perform a variety of tasks. Many thousands of interrelated and coordinated chemical reactions are generated in order to keep the organism alive, functioning, and able to repair damage.

Developmental processes exhibit a well-defined time and space order. Different events follow each other in an orderly time sequence, and at the same time spatial organization is seen in the embryo. Thus time and space coherence seems to be the rule in embryonic development. The birth of multicellular organisms comprises three aspects: *cell differentiation, pattern formation, and morphogenesis.*

13.1.1. Differentiation

A fertilized egg is subject to *cell cleavage*. It divides and multiplies into many thousands of cells. At the same time qualitative changes appear in the cell population. They transform, for example, into blood, skin, or liver cells. In another example, a cell from an embryonic chick limb bud differentiates into muscle or cartilage cells. The human organism harbors some two hundred different cell types. Two cell types have many *house keeping* proteins in common, but they differ in their *luxury* proteins, which characterize their specificity. For example, *hemoglobin* is a luxury protein which is responsible for the red color of the blood.

Cell differentiation results from a mother cell dividing into two daughter cells with alternative metabolic pathways. It is thought that these alternative pathways are the consequence of selective "on" or "off" switching of genes responsible for the synthesis of specific luxury proteins.

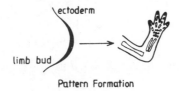

Pattern Formation

Fig. 13.1. Pattern formation. A characteristic patterned arrangement of skeletal elements emerges from chick limb bud. (After L. Wolpert, 1978.)

13.1.2. Pattern Formation

Differentiated cells organize themselves according to a well-defined pattern which is transmitted from one generation to the next. Each cell type has a well-defined spatial location. For example, from the chick limb bud, the cartilage, nerve, blood, skin, and muscle cells must differentiate and organize themselves in such a way that correct spatial and proportional relationships are held between the various cell types (see Fig. 13.1).

13.1.3. Morphogenesis

The phenomenon of morphogenesis is related to that of pattern formation. More precisely, it describes the mechanical processes by which organisms generate their specific form by the arrangement of their various tissues.

The processes leading to embryonic development are extremely complex and involve a great number of physical and chemical transformations which are mostly ill-defined at the present time. Thus the construction of a comprehensive mathematical model seems impossible. All we can hope for is to elaborate models based on well-known physico-chemical processes which describe a few fundamental events which occur in development: the spontaneous onset of spatial inhomogeneity in otherwise uniform morphogenetic fields, and the probable mechanism of cell differentiation in such a newly created assymmetric environment.

The present chapter examines the modeling of cellular differentiation and pattern formation. However, following the customary misuse of the word, we shall often speak of these two combined processes as morphogenesis, although mechanical effects will not be considered. The aim of theories of morphogenesis is to find universal principles underlying developmental processes in multicellular embryos, based on a small number of rules governing cell behavior.

Let us recall that the reaction-diffusion equations were used for the first time in biology by Turing (1952) precisely in the context of the emergence of patterns in a homogeneous tissue. Today we have much more elaborate models of pattern formation and morphogenesis; one of these models is summarized in the following sections.

Theories of developmental events usually do not rely on well-defined biochemical processes, as was the case for metabolic oscillations and slime mold aggregation. The systems under consideration are extremely complex, and involve the successive unfolding of many ill-defined chemical processes. To make matters worse, in morphogenetic systems the biochemical processes are not the same from one point of the system to the next. And since the organisms are extremely complex, theoretical predictions cannot be tested in an unambiguous manner. These facts render the theories of developmental processes more speculative and at the same time more desirable. Presently we are far from a satisfactory reductionist knowledge of developmental processes. Even if a complete biochemical description of the embryo becomes possible some day, it will not lead to a total understanding of developmental problems. Biochemical studies in general throw little light on the dynamics and coherent behavior of cellular assemblies. Thus they ignore the resulting cellular cooperation which is the essence of pattern specification.

Before we go any further, we must briefly summarize the embryonic development of a multicellular organism.

13.2. DEVELOPMENT OF THE EMBRYO

Heterosexual reproduction necessitates the formation of eggs and spermatozoids. The latter are minute, whereas eggs are usually large. The egg contains reserve material for the future use of the developing embryo. It contains different fats, yolk proteins, and various other substances which give such a great nutritive value to an omelet.

The egg of the female and the spermatozoid of the male are the only cells which contain one half less genetic material than the remaining cells of the organism. When an egg and a spermatozoid combine into a unique fertilized cell, the normal amount of the genetic material of the species is recovered. After fertilization, the egg divides and gives birth to two cells called *blastomeres*. Each one of these cells divides in turn into two still smaller cells. Another cell division produces an eight-cell embryo. The cell cleavage continues, as smaller cells are formed. At some point, the cell population is no longer homogeneous and differentiates into several cell types. After complex and intensive cell movement, a new structure called the *gastrula* appears in the embryo. The gastrula is formed from distinct sheets of cells that surround an inner cavity (see Fig. 13.2). From here on, each embryo follows its own morphogenetic program.

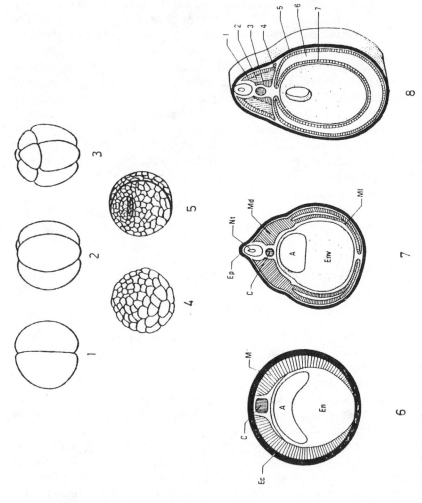

Fig. 13.2. Early embryo of amphioxus. (4–5) Early cleavages form the blastula. (6–8) Early and late gastrula. (From P. Van Gansen, *Abrégé de Biologie Générale* © Masson 1984.)

All experimental data converges on the fact that the genetic information is transmitted in its integrity to all progeny of the fertilized egg. A skin or blood cell both possess the same genetic information, although different parts of the genome are expressed in each cell type. Therefore, embryonic development may be regarded as a programmed, sequential, gene expression.

Little is known about the molecular mechanism of the "on" and "off" switches of genes in higher organisms. The current understanding is that the mechanism may be broadly similar to the bacterial induction model of Jacob-Monod shown in Fig. 11.13. However, we must keep in mind that some mammalian cells may contain 800 times more DNA than *E. coli*, and can instruct the synthesis of a great number of different types of proteins.

We may assume that, in multicellular organisms, instead of an external sugar source the cells use internal chemical signals which are embedded in the cytoplasm. These cytoplasmic factors are assumed to be distributed inhomogeneously in an otherwise identical cell population. As we have seen in Section 11.7, enzyme induction may be coherent and exhibit multiple steady states when a threshold value of inducer is reached. The cells start synthesizing new enzymes and proteins not existing in the subthreshold cells. Thus two different cell types are created from a single cell.

A human fertilized egg differentiates into about 200 cell types. The aim of the present chapter is not to account for such diversity; rather, we hope to find general principles which may govern cell differentiation and pattern formation.

13.3. DESCRIPTIVE THEORIES OF DEVELOPMENT

Relying on experimental data, embryologists have elaborated three different theories to describe developmental events. These are mosaic theory, reference point theory, and noninstructive theory. Each hypothesis best fits the experimental data of a given embryo.

13.3.1 Mosaic Theory

This theory assumes that instructions for the development of organisms are engraved as a mosaic pattern in cell membranes and cytoplasms. During cell division each cell group will receive a particular assignment.

13.3.2. Reference Point Theory

In this theory, it is assumed that several "reference" or "source" points exist in the egg from which substances called *morphogens* diffuse and form appropriate gradients. Thus, cell cleavage produces units with different morphogenetic content. Therefore, cell nuclei are surrounded by inhomogeneous cytoplasmic factors. This fact may bring about differential gene expression.

13.3.3. Noninstructive Theory

In this theory, the membrane and cytoplasm of the egg do not contain instructions for the new nuclei which are formed during cleavage. Reference points or gradients appear in the multicellular embryo after cell cleavage and cellular communication. Many such embryos have been studied experimentally.

Most mammalian eggs, for example, the mouse or rabbit oocyte, are completely homogeneous; no mosaic patterns or reference points have been demonstrated for these cells. A fertilized mouse egg may develop even in vitro or in a simple chemical medium up to an embryo formed of 64 to 128 cells. If this embryo is transferred into the testis or brain of the male, it will continue its development into a fetus. Already two types of cell population may be identified inside the 64-cell embryo. The outside cells, called trophoblasts, are functionally and biochemically different from the inside cells. The subsequent developmental fates of the two cell types are also not similar. This fact indicates the emergence of biochemical inhomogeneity inside the previously homogeneous cells of the embryo. The time of the onset of this first gradient may be determined by appropriate experimental procedures.

If the blastomeres of a two-cell mouse embryo are separated, each cell can form a whole mouse. Had they not been separated, only one mouse would develop. Therefore, an isolated blastomere has acquired a higher potential for development than the attached cells. This phenomenon is known as *embryonic regulation*.

Isolated blastomeres from a 4- or 8-cell stage may each form a primitive embryo. A single blastomere of an 8-stage rabbit embryo may form a whole rabbit. Therefore, we may suspect that in the rabbit embryo, irreversible cellular differentiation arises sometime after 8 cleavages.

In another set of experiments with mouse embryos, it is shown that cell position can markedly alter cell fate. An 8-stage embryo is dissociated into single blastomeres. The latter surround a normal embryo. The progeny of the outer cells contribute only to the formation of the trophoblast and yolk sac of the composite embryo, whereas normally they would have formed a complete animal. Data from mouse embryo experiments may be summarized as follows:

1. Cellular differentiation appears in the embryo consequent to cell cleavage and communication.
2. The developmental fate of the cells of an early-cleaving egg is

determined by their position in the embryo. Their fate may be altered by changing their position.

13.4. THE CONCEPT OF POSITIONAL INFORMATION

Let us first introduce the important concept of morphogenetic fields defined as an ensemble of functionally coupled cells governed by the same regulatory processes. A typical morphogenetic field has less than 100 cells. For example, in sea urchin gastrula a field of 30 cells measures 0.2 millimeters.

The idea that the fate of a cell in a developing embryo is largely specified by its position within the morphogenetic field was first stated by Driesch in 1890. The same idea was implicit in the theories of Child when in 1920 he stated that physiological gradients were responsible for pattern specification in the development of species. In 1969 these phenomenological concepts were revived and formalized by L. Wolpert and took the name of *positional information*.

The concept of positional information suggests that the positions of cells are specified within the morphogenetic field with respect to various reference points. Each cell genome "reads" its positional information within the field and differentiates accordingly. Therefore, differential gene activation is spatially determined by various threshold values. Essential features of the model may be seen in the following unrealistic example.

Consider a high mountain with its base in sea water. The concentration of the oxygen of the air decreases regularly from sea level to the mountain top. Now imagine a handful of *identical* flower seeds, such that the fate of their germination depends crucially on the oxygen concentration of the air. For example, at the mountain top they produce yellow flowers, whereas at sea level the flowers are blue. Many other colors arise in intermediary altitudes. Therefore, when a seed falls at a given altitude its fate is determined solely by the oxygen content at that level and is totally independent of the fate of the neighboring flower fields.

This unrealistic example raises three important questions that must be answered in the framework of any coherent theory of development:

1. Why is there a source and a sink value of oxygen (morphogen)?
2. What is the molecular mechanism underlying the gradient formation?
3. How does the unique seed genome respond differentially to the oxygen concentration?

The first two questions, transposed into morphogenetic fields, require the elaboration of a theory that accounts for the generation of positional information inside a homogenous cell assembly. Many such theories do exist. They may be classified into two distinct categories. In a first group of models one assumes the existence of a source and a sink at the field boundaries. Different mechanisms are given for the establishment of a gradient between these two boundaries. In other models, in a field of identical cells sources, sinks and gradients are established spontaneously and simultaneously by a unique mechanism. As we are mainly interested in self-organizing systems, more emphasis is given to the second group of models. The third question is the subject matter of *positional differentiation*.

13.5. SPONTANEOUS GENERATION OF POLARITY IN MULTICELLULAR SYSTEMS

Assuming that the gradient theory and the concept of positional information reflects the reality of developmental events, we must still answer the fundamental question; how are gradients generated in morphogenetic fields of seemingly identical cells? In other words, we must explain how sources and sinks appear in the boundaries of the morphogenetic fields.

Let us consider a morphogenetic field of N cells and assume that two morphogens travel the field by some permeation mechanism, for example, by tunneling through the gap junctions. Moreover, the two morphogens react together according to some specified nonlinear kinetics. Now if abstraction is made of finite barriers formed by cell membranes, a morphogenetic field may be considered as an ordinary reaction-diffusion system such as the one described in Section 9.8. Moreover, we showed there that, if morphogens do not leak across the boundaries of the field, then the homogeneous state of a reaction-diffusion system may bifurcate into various inhomogeneous solutions. In particular, a morphogenetic gradient in the shape of a polar structure may be obtained as a first bifurcating solution.

However, it is more desirable to demonstrate the onset of polarity in a morphogenetic field in which cell membranes together with all their complex intercellular communication devices are considered.

13.5.1. The Discrete Model

From the outset we realize that a piece of biological tissue with its hundreds of compartmentalized cells is essentially different from a chemical vessel of the same size containing a homogeneous mixture. Moreover, cell communication is mediated by highly nonlinear biochemical processes, such as pumps, channels, receptors, and gap junctions. Therefore, we need to write a pair of dynamic equations for each cell, and also express the passage of chemicals from one cell to the next.

Fig. 13.3. A one-dimensional morphogenetic field. Each cell exchanges morphogens only with its first neighbors. Zero-flux boundary conditions are assumed.

The formulation that follows may be easily extended to the actual three-dimensional tissues. However, the ideas and the results may be seen more simply with one-dimensional systems. Such one-dimensional organisms are seen in nature. For example, a blue-green algae is a long filament of one-cell thickness. The cells are differentiated into vegetative cells and "heterocytes." These simple and extremely useful organisms are cultured to produce energy from biomass. They live and multiply in sewers and at the same time clean the water for future use.

Let us go back to our model construction and consider a multicellular system composed of N identical cells (see Fig. 13.3). In each cell there are m reacting chemical substances which interact according to kinetic laws that need not be specified at this stage and are noted by a function $f^l(X_i^1,\ldots,X_i^m)$. The concentration X_i^l of substance l in cell i changes in time for two reasons: X_i^l is consumed or produced inside the cell i, or is modified by losses or gains through interaction with its first neighbors, cells $i+1$ and $i-1$. The complex intercellular communication processes are represented by $g^l(X_{i+1})$ and $g^l(X_{i-1})$; therefore, the most general kinetic equations for a one-dimensional morphogenetic field of N cells are:

$$\frac{dX_i^l}{dt} = f^l(X_i^l \cdots X_i^m) + g^l(X_{i+1}) + g^l(X_{i-1}),$$

$$l = 1,\ldots,m,$$

$$i = 1,\ldots,N; \tag{13.1}$$

The *contact functions* g^l are usually highly nonlinear. However, if molecules cross cell boundaries via gap junctions, it is reasonable to think that the number of X_{i+1}^l molecules that leave cell $i+1$ to enter cell i, to a first approximation, is linearly proportional to the concentration of X_{i+1}^l in cell $i+1$. The same is true for the $i-1$ neighbor. If the proportionality coefficient D describes the permeability of the membranes, and assuming the same permeability for all cells, the contact function for cell i takes the simple form

$$g^l = D[X_{i+1}^l + X_{i-1}^l - 2X_i^l] \tag{13.2}$$

Here $-2DX_i^l$ describes the loss of X_i^l from cell i.

Moreover, if we assume that, in each cell, two morphogens, $X_i^1 = X, X_i^2 = Y$, are produced, for example, via the Brusselator scheme, then the morphogenetic field of N cells is described by the following equations:

$$\frac{dX_i}{dt} = A + X_i^2 Y_i - (B+1)X_i + D_X[X_{i+1} + X_{i-1} - 2X_i]$$

$$\frac{dY_i}{dt} = BX_i - X_i^2 Y_i + D_Y[Y_{i+1} + Y_{i-1} - 2Y_i], \qquad i = 1,\dots,N \qquad (13.3)$$

One verifies immediately that $X = A$, $Y = B/A$ are the unique homogeneous steady states of Eq. (13.3).

Let us introduce a small perturbation in one of the cells, $X_i = A + x_i$, and follow the time evolution of morphogen distribution throughout the field of N cells. Moreover, let us assume that the system is under zero-flux boundary conditions. The linear stability analysis of Eq. (13.3) may be performed by the usual methods of Section 9.7. However, at present the characteristic equation is a polynomial of $2N$ degree. Fortunately, thanks to the symmetry of intercellular interactions, one shows that, if cells are arranged in a closed ring, then

$$D[X_{i+1} + X_{i-1} - 2X_i] = 2D\left[1 - \cos\frac{2\pi k}{N}\right]X_i, \qquad k = 1,\dots,N$$

With the help of this relation, the characteristic equation reduces to N second-degree identical polynomials as given by 9.28. However, here the permeability coefficient multiplies a factor which is a function of cell number N.

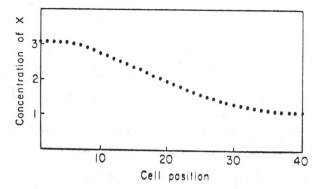

Fig. 13.4. Spontaneous generation of polarity in an array of 40 homogeneous cells in contact and subject to zero-flux boundary conditions.

The standard linear stability analysis shows that, for example, if $A = 2$, $B = 4.6$, $D_X = 266$, and $D_Y = 1330$, the steady state is unstable and the singular point is a saddle point. The direct numerical integration of Eq. (13.3) shows the spontaneous onset of a morphogenetic gradient in the field of N cells (A. Babloyantz 1977). Some cells contain more morphogen than the steady state, while other cells synthesize less morphogen (see Fig. 13.4).

Spontaneous pattern formation is not restricted to models with diffusion-type processes. Instabilities may arise in systems with complex intercellular communications. It can be shown easily that if we choose $D = (N/l)^2 D'$, the set of $2N$ Eq. (13.3) are equivalent to a reaction-diffusion mechanism where the transport phenomenon is approximated by the discrete representation of Fick's law of diffusion. In this expression l is the total length of the organism, N the total cell number, and D' the Fick diffusion coefficient. If the morphogenetic field comprises a large number of cells, $N \to \infty$, one recovers exactly the reaction-diffusion equations of Section 9.3.

13.5.2. The Continuous Model

In the description of bifurcating solutions of morphogenetic fields, the length l of the system is the most natural bifurcation parameter. To show this point in a simple manner, we go back to the continuous description of reaction-diffusion mechanisms. As we have already seen, this approach is legitimate for large N and linear intercellular communication processes.

We showed that analytical as well as computer results have proven unambiguously the spontaneous emergence of patterns in a wide class of reaction-diffusion systems. The stability of the first bifurcating solutions with respect to small fluctuations has been shown by analytical methods. Therefore, a theoretical approach to the problem of morphogenesis leads to the concept of dissipative structures, and more particularly to those bifurcations which produce spatially inhomogeneous solutions in the reaction-diffusion systems.

Many different two-variable functions may produce spontaneous pattern formation; however, from a morphogenetic point of view they may be classified into two categories:

1. The two morphogenetic gradients of X and Y, resulting from a first bifurcation, have their high and low points at the same extremities of the field. Therefore the two gradients run parallel.
2. The X and Y gradients run in opposite directions. (The Brusselator scheme belongs to this second category.)

Two parallel gradients may be found in the following scheme:

$$\xrightarrow{k_1 A} X$$

$$C \xrightarrow{k_2 X^2} Y$$

$$X \xrightarrow{k_3} F$$

$$Y \xrightarrow{k_4} G$$

$$Y + P \underset{k_6}{\overset{k_5}{\rightleftharpoons}} X$$

$$B + X + P \xrightarrow{k_7} P + 2X \tag{13.4}$$

A morphogen X is produced at a constant rate $k_1 A$ as well as via an autocatalytic step that requires a third molecule P. X catalyzes the formation of Y from a precursor C at a rate $k_2 C X^2$. Moreover, P transforms Y into X. Finally, the intermediates X and Y have finite lifetimes; they are either used up or degraded by the cells and leave the reaction medium via unimolecular steps.

If a quasi-steady state assumption is made for the substance P, the scheme (13.4) may be described by two differential equations which were postulated by Gierer and Meinhardt for description of morphogenetic fields:

$$\frac{\partial X}{\partial t} = k_1 A + \frac{k_6 k_7 B X^2}{k_5 Y} - k_3 X + D_X \nabla^2 X$$

$$\frac{\partial Y}{\partial t} = k_2 C X^2 - k_4 Y + D_Y \nabla^2 Y \tag{13.5}$$

As the two morphogens must not flow out of the boundaries of the developing field, zero-flux boundary conditions are assumed.

$$\left(\frac{\partial X}{\partial r}\right)_0 = \left(\frac{\partial X}{\partial r}\right)_l = \left(\frac{\partial Y}{\partial r}\right)_0 = \left(\frac{\partial Y}{\partial r}\right)_l = 0 \tag{13.6}$$

Here l is the length of the field.

As usual, we must evaluate the uniform steady states X^s and Y^s and test their stability by introducing inhomogeneous perturbations $x = x(r, t)$ or $y = y(r, t)$ in the system.

Following A. Babloyantz and J. Hiernaux we illustrate the spontaneous onset of a morphogenetic gradient in reaction-diffusion systems with the help of the simple scheme (13.4). However, let us keep in mind that the final conclusions are model-independent and are applicable to a large class of reaction-diffusion systems.

The perturbations x and y must satisfy the same boundary conditions as Eq. (13.6). Therefore, functions $x = c_1 e^{\omega_m t} \cos(m\pi r/l)$ and $y = c_2 e^{\omega_m t} \cos(m\pi r/l)$ with $m = 0, 1, \ldots, N$ are appropriate solutions to the linearized equations, since their derivatives at the boundaries, $[ce^{\omega m t} \sin(m\pi r/l)]_0^l = 0$, vanish. Following our usual method of analysis, we must evaluate the roots of the characteristic equation

$$
\omega^2 + \omega \left[k_3 + k_4 - \frac{2X^s k_7 B}{Y^s} + D_X \frac{m^2 \pi^2}{l^2} + D_Y \frac{m^2 \pi^2}{l^2} \right]
$$

$$
+ \frac{m^4 \pi^4 D_X D_Y}{l^4} + \frac{m^2 \pi^2}{l^2} \left(k_3 D_Y + k_4 D_X - \frac{2X^s k_1 B D_Y}{Y^s} \right)
$$

$$
+ k_3 k_4 - \frac{2X^s k_4 k_7 B}{Y^s} + \frac{2(X^s)^3 k_2 k_7 BC}{(Y^s)^2} = 0 \tag{13.7}
$$

In this equation, the parameters which characterize the multicellular aspect of the system are the average cellular permeability coefficients D_X, D_Y, and the field length l. The integer m determines the particular shape of spatial patterns.

From linear stability analysis we may predict that inhomogeneous patterns may arise if the homogeneous steady state is unstable and exhibits a saddle-type singular point. The latter may appear if the independent term of the characteristic equation changes sign toward the negative values (see Section 9.7).

The analysis of the characteristic equation shows that the independent term Δ is a function of chemical rate constants, diffusion constants, the field dimension l, and the wave number m. Figure 13.5 shows the plot of independent term Δ as a function of the length of the field for several values of m.

We see that for $m = 1$, Δ vanishes at l_1^1, indicating a possibility for the onset of instabilities. If the total length of the system is less than l_1^1, the homogeneous steady state of the system remains stable and there is no possibility of source and sink formation. However, if the length is increased beyond l_1^1, the singular point is a saddle point and inhomogeneous patterns may appear spontaneously in the morphogenetic field.

Moreover, we can guess from Fig. 13.5 that up to the length l_2^1, one may only obtain a structure with one high and one low point. This may be seen, for example, by examining explicitly the perturbation function $\cos(\pi r/l)$ and remembering that the final solution is $X = X^s + ce^{\omega t} \cos(\pi r/l)$. If, on the other hand, we take a system of length l such that $l > l_1^1$ and less than l_3^1, two possibilities arise. Because of the overlapping of values of m for a given length of the system, structures with one high region and others with two high regions are possible. The form of the emerging pattern is a function of the magnitude of the initial perturbations and of their location in the system.

Once the bifurcating length of the system has been determined with the help of linear stability analysis, the above conjectures may be verified by direct

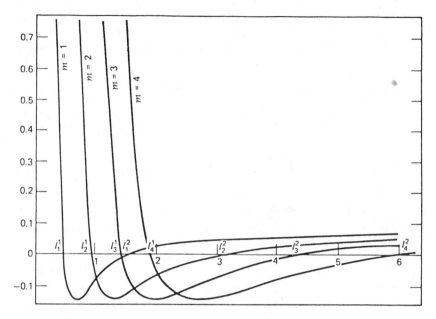

Fig. 13.5. Variations of characteristic determinant, computed from Eq. (13.7), as function of the length l for different values of wave number m, showing wavelengths that can be fitted within a given length.

numerical integration of Eq. (13.5) or by construction of approximate analytical solutions. Predicted patterns are observed, and, in particular, if the length of the system is between l_1^1 and l_2^1, one finds typical polar solutions, such as the one depicted in Fig. 13.6. This structure displays a gradient which appears spontaneously in the morphogenetic field if random fluctuations destabilize the homogeneous state.

A most important point is that we have a theorem demonstrating the stability of this first bifurcating solution. This is a crucial property, since in the present context only stable gradients are appropriate for the generation of positional information.

As l increases further, the system may become unstable for higher values of m. Thus a large number of wavelengths can be fitted into the system and more complex patterns with several hills and valleys may appear.

In the Gierer-Meinhardt model, Eq. (13.5), substance X stimulates its own production and at the same time produces the morphogen Y. Therefore, it is natural that both morphogens have their maximum in the same location. On the other hand, Y diffuses faster than X, $D_Y > D_X$, and at the same time inhibits the production of this morphogen. The combination of these two factors is responsible for the sharpness of X and flatness of Y patterns.

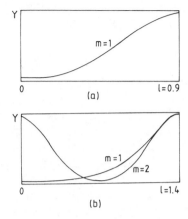

Fig. 13.6. (a) If $l > l_c$, the first polar structure appears in a reaction-diffusion governed morphogenetic field. (b) For larger l, two wavelengths $m = 1$ and $m = 2$ may appear in the same field.

In the example of reaction-diffusion equations based on the Brusselator scheme, Eq. (9.2), the system becomes unstable as a chemical parameter B exceeds a threshold value B_c. In the present example the parameter which brings about bifurcations is a geometrical factor. This is in agreement with the fact that in the sequence of developmental processes, change in size is certainly one of the most conspicuous events.

Reaction-diffusion mechanisms provide positional information, a sort of prepattern for the developmental fields. This information must be encoded and translated by the individual cells into a secondary pattern, the one visible to the naked eye. The concept of *positional differentiation* designates this secondary pattern formation which involves the orderly expression of genes.

Models accounting for positional differentiation may be constructed by combining reaction-diffusion and induction mechanisms. In these models, once the polarity is generated in the embryo, the environment of the identical cells is no longer uniform. Following a mechanism such as the one seen in Section 11.7, some cells may be induced to synthesize specific proteins, thus creating a two-type cell pattern.

13.6. SCOPE AND LIMITATIONS OF THEORIES OF MORPHOGENESIS

The mathematical theory of morphogenesis developed in the present chapter relies on the following assumptions:

1. There exist at least two reacting morphogens which travel throughout the morphogenetic field.
2. Each pattern is characterized by a well-defined geometry of the field.

Although the existence of morphogens and morphogenetic gradients has been postulated for several decades, as yet their biochemical nature has not been unambiguously demonstrated. The reason for this fact is that many unspecific agents or organic salts initiate morphogenesis. In recent years, H. C. Shaller has isolated four substances from *Hydra* that she claims are foot inhibitor, foot inducer, head inducer, and head inhibitor morphogens.

Hydra is a small, fresh-water inhabitant which measures about one centimeter. Under normal circumstances it reproduces asexually by forming buds in well-defined regions of the body. Hydra is a simple organism formed from a body column, surmounted by a head, mouth, and tentacles at one end and a sticky foot at the other end. The body comprises only few cell types. This simple organism has been a choice experimental material of embryologists for over seventy years.

The animal has an extraordinary power of regeneration. If a piece is cut from the gastric region, a new head and foot regenerate and a small, complete, new animal is formed. Moreover, the polarity of the gastric piece is conserved and is that of the intact animal. The system regulates without growth and the phenomenon is known as *morphalaxis*. Many other transplantation experiments may be performed with hydra. For example, different pieces from different parts of the organism of a donor animal are grafted onto the body of a host. The grafted piece may act as organizer and transform the otherwise already committed cells of the host organism into a head or foot. For example, a piece from the head region grafted in the gastric region induces the formation of a new head and inhibits the formation of other parts (see Fig. 13.7).

The head inducing capacity of the grafted piece decreases when pieces are taken from a lower position in the body column of the donor animal. These facts may be interpreted as a proof of the existence of a head inducing biochemical gradient with its maximum at head region and minimum in the foot area. Moreover, a piece from the foot region of the donor, if transplanted in the gastric region of the host, induces foot formation. Therefore, the existence of a foot inducing gradient could be shown by testing the foot inducing capacity of different parts of the donor animal. As might be expected, this gradient is the opposite of the head inducing gradient and is maximum at foot level.

In another series of experiments, a piece from the head region of the donor was grafted to different regions of the body of the host. It was found that these organizers could not induce a new head in the host if the implant position was

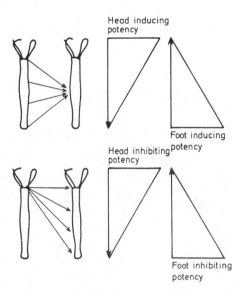

Fig. 13.7. Schematic representation of head- and foot-inducing potency in some transplantation experiments. At the top the position in the donor from which the implant was taken is varied, at the bottom the position in the host is varied. (From C. J. Grimmilkihyen *et al.* 1979, reprinted by permission of Elsevier Publications (Cambridge).)

too close to the head region. This fact shows the existence of a head inhibiting gradient. A foot inhibiting gradient was shown in a similar series of experiments.

Assuming the existence of morphogens, Gierer and Meinhardt made a correspondence between the various bifurcating solutions of the reaction-diffusion equation (13.5), computed in the two-dimensional morphogenetic fields, and the grafting experiments in hydra. These experiments may be understood quite satisfactorily in the framework of dissipative structures.

The theories of morphogenesis outlined in the present chapter are quite adequate for the description of pattern formation in the presence of growth and differentiation. For example, in Section 13.5 we showed that if the length of a morphogenetic field exceeds a critical value, a region of high concentration of activator appears. If one identifies this high concentration area with the head region of hydra, then one obtains an explanation for an experimentally well-known fact, namely, that a uniform piece from the hydra must exceed a critical size before regenerating a new animal.

Another choice material for experimental and theoretical considerations in embryology is the *Drosophila*, a small fly which lives on decaying fruit. In this fly the development of the wings from the embryo requires a change in size,

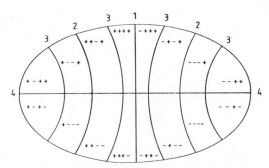

Fig. 13.8. Schematic representation of a developing Drosophila egg. Successive compartmental lines are predicted from successive binary choices. (Reprinted by permission from S. A. Kauffman et al., *Science*, Vol. 199, 1978, p. 259. Copyright © 1978 by the AAAS.)

shape, composition, and is followed by compartmentalization. Reaction-diffusion mechanisms have been used by S. Kaufmann to account for the succession of these events in wing development. One argues that, as the wing tissue increases in size, new bifurcating solutions arise in the system which exhibit more and more maxima and minima in concentrations. The reaction-diffusion mechanism is also applied to the developing egg in the absence of size increase. Here decrease in cytoplasmic viscosity (change of diffusion coefficient), considered as a bifurcation parameter, selects successively different chemical patterns. As D_X and D_Y are increased, patterns with many peaks and valleys appear and define compartmental lines (see Fig. 13.8).

13.7. PATTERN REGULATION

In the previous section we saw how an isolated piece of a gastric region of a hydra gives rise to a small but otherwise normal animal. Apparently there is no growth in the isolated gastric piece, and regeneration is the result of cellular reorganization.

In Section 13.3 we described the phenomenon of embryonic regulation in cleaving eggs by showing that 1/8 of a rabbit embryo gave birth to a whole rabbit. Quite generally, the sequence of development can be brought to normal completion sometimes even after more or less severe disturbances, such as addition, deletion, and tissue rearrangement. This capacity for regulation relies entirely on cellular communication processes.

Transposed into the framework of reaction-diffusion theory, embryonic regulation requires that the solution to the pertinent differential equations remain invariant if the size of the system is reduced. In other words, we require that reaction-diffusion patterns exhibit a *size invariance* or *pattern regulation* property. This property is defined as follows: if a part of a developing field is removed and the lost boundary is restored at the cut surface, the remaining part regulates and restores the original pattern in a manner such that the respective proportions of the different parts of the pattern are independent of the total size of the organism. For example, if in normal hydra the head forms 1/5 of the total length of the animal, we require that this proportion remain the same in the regenerating small piece.

The property of size invariance implies that if a system of length l has a concentration m of morphogen at point r, and if the size of the system is reduced to l', there must be a point r' in the new system with the same concentration m as before, and such that $r'/l' = r/l$.

The property can only be demonstrated rigorously if a morphogen diffuses between two imposed source and sink values S_1 and S_2 without participating in any biochemical reaction. In this case the steady-state solution for morphogen distribution along the field is found easily and is given by

$$m = \frac{S_1 - S_2}{l} r + S_2 \tag{13.8}$$

This expression satisfies the condition for size invariance stated above, and is independent of the value of the diffusion coefficient of the morphogen.

Unfortunately, size invariance cannot be demonstrated in morphogenetic fields in which gradients are the consequence of reaction-diffusion interactions and do not result from imposed boundary values. The reason for such a failure is that the linear diffusion process, by its intrinsic nature, is size-dependent. Consequently, the patterns it generates are always size-dependent, and therefore are nonregulatory.

One can show easily that size invariance is only possible if a new constraint is imposed on the system, such as, for example, if the field length is shortened then the morphogen must diffuse more slowly,

$$\frac{D}{l^2} = \frac{D'}{l'^2} \tag{13.9}$$

As yet there are no experimental data suggesting the possibility of such a large permeability change in the regulating morphogenetic fields.

The failure of reaction-diffusion models to account for size invariance may be attributed to the fact that cell interactions have been approximated by an average diffusion process. In this approach the entire morphogenetic field is treated as a unique chemical vessel, and cell individuality, cell membranes, and the number of intercellular contacts are ignored.

Let us go back to the discontinuous description of morphogenetic fields and consider individual cells and their immediate neighbors. In a two- or three-dimensional tissue the number of the first neighbors of a given cell may vary considerably. We show in the next chapter that this number may become a bifurcation parameter. That is, for a given metabolism and identical permeability coefficients, the self-organizational properties of the tissue are determined by the number of cell-cell contacts.

Now let us assume that in a three-dimensional morphogenetic field each cell is in contact with twelve immediate neighbors. Moreover, reaction-diffusion processes establish morphogenetic patterns in the field. One shows, by constructing approximate analytical solutions, that if a part of the cells of the morphogenetic field is discarded, the remaining cells may show pattern regulation by simply lowering the number of intercellular contacts. This fact may be achieved by local cellular motion or cell membrane contraction.

While the cell-contact number changes, two situations may be considered:

1. Cell volume remains constant. In this case, if the number of contacts are reduced to six neighbors, and if N is the number of cells of the larger field and N' that of the reduced tissue, then one shows that size invariance is possible provided $N = 1.19 N'$.

2. The number of cell-contacts changes while cell surface remains constant. In this case $N = 1.41 N'$ ensures pattern regulation.

Size invariance may also be shown in tissues where the number of intercellular contacts remains identical in the injured field provided cells decrease the surface area in contact with neighboring cells, thus creating intercellular space. Such a phenomenon may be the result of cytogel contractibility.

In regenerating morphogenetic fields, probably both mechanisms operate simultaneously. It must be noted that these results are general and model-independent. They can be applied to any multi-unit ensemble exhibiting coherent properties which characterize the system as a whole.

*   *

*

In the next chapter we develop a more general formalism for the study of multiply connected, multi-unit ensembles of cells. We show how the cell-cell contact number, in an arbitrarily densely connected network, becomes a bifurcation parameter. We show the relevance of such ideas for the study of neural networks.

Chapter 14

SELF-ORGANIZATION IN MULTIPLE-UNIT SYSTEMS

14.1. INTRODUCTION

In Chapter 13 we extended the concept of dissipative structures to the developmental fields made of a large collection of interconnected reaction-diffusion units. By considering only linear contact functions and first-neighbor interactions, we showed that large developmental fields may be approximated quite satisfactorily by a single reaction-diffusion system. Many fields of study, ranging from biology to the social sciences, focus on a wide variety of systems, made of a large number of units interconnected by highly nonlinear contact functions. In such ensembles a unit may be connected to many or all other units of the system.

Our central nervous system is a good example of such a complex organization. Each cell of the brain makes a few thousand contacts with the cells of the same tissue. It is the cooperation of the ensemble of cells that generates the unique properties of the nervous system.

The cooperation between units which creates a new collective behavior exists also at the level of organisms. For example, in many insect societies there is a well-defined social order and collective behavior. The individuals create this behavior by a complex intercommunication network. In some ant societies a chemical substance called pheromone is synthesized by individual ants and serves as a tracer for conveying information about nutrient resources to their fellow ants, thus organizing the collective behavior of food gathering. The social and economic structures of our societies are other examples of multi-unit interconnected ensembles. The behavior of these systems cannot be understood by considering only unit properties.

It seems desirable, therefore, to extend the concept of self-organization and the methods of investigation of dissipative structures to the multi-unit and multiply connected systems.

14.2. SELF-ORGANIZATION IN MULTIPLE-UNIT SYSTEMS

Multiply-connected ensembles may be investigated by selecting appropriate variables for the description of time change of unit properties, and by relating the units together with appropriate contact functions. If a single unit may be described by two variables, one is led again to the construction of differential equations similar to Eq. (13.1),

$$\frac{dX_i}{dt} = f(X_i, Y_i) + g_X(X_{i-n}, \ldots, X_{i+n}, Y_{i-n}, \ldots, Y_{i+n})$$

$$\frac{dY_i}{dt} = h(X_i, Y_i) + g_Y(X_{i-n}, \ldots, X_{i+n}, Y_{i-n}, \ldots, Y_{i+n}) \qquad (14.1)$$

However, the nonlinear operator g may now contain as many variables as there are units in the ensemble. Here the integer n designates the *connectivity number*, and $2n$ is the total number of connections made by a unit with other members of the network.

In general, homogeneous steady states of the set of Eq. (14.1) must be evaluated by numerical methods. The stability of these solutions and the various unstable points may be found by the usual methods of linear stability analysis. In the present situation, however, one must also linearize the nonlinear contact operator g by developing it in a Taylor series around the steady states.

In order to see in a simple manner the influence of the connectivity number upon unstable solutions of Eq. (14.1), we consider a linear contact operator $C[1/n(X_{i-n} + \cdots + X_{i-1} + X_{i+1} + \cdots + X_{i+n}) - 2X_i]$. The factor $1/n$ ensures that the homogeneous steady state is identical for every connectivity number. Let us note that $n = 1$ corresponds to the first-neighbor interactions of Section 13.6.

The linearized equations are

$$\frac{\partial x_i}{\partial t} = \left(\frac{\partial f}{\partial X}\right)_s x_i + \left(\frac{\partial f}{\partial Y}\right)_s y_i$$

$$+ C_X\left[\frac{1}{n}(x_{i-n} + \cdots + x_{i-1} + x_{i+1} + \cdots + x_{i+n}) - 2x_i\right]$$

$$\frac{\partial y_i}{\partial t} = \left(\frac{\partial h}{\partial X}\right)_s x_i + \left(\frac{\partial h}{\partial Y}\right)_s y_i$$

$$+ C_Y\left[\frac{1}{n'}(y_{i-n} + \cdots + y_{i-1} + y_{i+1} + \cdots + y_{i+n}) - 2y_i\right]$$

$$i = 1, \ldots, N \qquad (14.2)$$

The characteristic equation is a polynomial of the form

$$\omega^{2N} + \mathscr{A}\omega^{2N-1} + \cdots + Q\omega + R = 0 \tag{14.3}$$

For large N analytical discussion of the behavior of the $2N$ roots of this equation becomes impossible. The problem, however, may be reduced to the discussion of a second-degree equation if various units may be decoupled from each other by some mathematical artifice.

One can show rigorously that such a decoupling is possible for the following unit arrangement. The ensemble of N cells form a closed ring, therefore the system is subject to periodic boundary conditions. Moreover, a given unit must have the same number of connections in both sides, as shown in Fig. 14.1. In this case it is possible to find a set of functions $x_1, \ldots, x_N$ such that

$$\frac{1}{n}[x_{i-n} + \cdots + x_{i-1} + x_{i+1} + \cdots + x_{i+n}] = \lambda^k x_i \tag{14.4}$$

with

$$\lambda^k = 2\left[1 - \frac{1}{n}\frac{\cos\left((n+1)(\pi k/N)\right)\sin\left(n\pi k/N\right)}{\sin\left(\pi k/N\right)}\right], \qquad k = 1, \ldots, N$$

The constants λ^k are the *eigenvalues* and $x_1, \ldots, x_N$ are the *eigenfunctions* of the linearized contact operator.

If Eq. (14.4) is introduced into the linearized Eqs. (14.2), the latter decouple into N identical pairs of equations. Thus, with the help of a mathematical property the study of a multi-unit ensemble is reduced to the investigation of

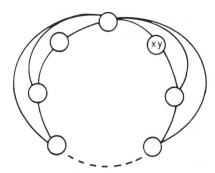

Fig. 14.1. Some elements of a network subject to periodic boundary conditions. Each element is connected to the same number of units on both sides.

the familiar two-variable characteristic equation

$$\omega^2 + \omega \mathscr{F}(C_X, C_Y, \lambda_x^k, \lambda_y^k) + \mathscr{B}(C_X, C_Y, \lambda_x^k, \lambda_y^k) = 0 \qquad (14.5)$$

The new feature of Eq. (14.5) is that the eigenvalue λ^k which expresses the influence of all neighbors on cell i is an explicit function of the connectivity number n. Figure 14.2 shows the first four values of λ^k as a function of n for a 40-unit ensemble. It is seen that eigenvalues increase with the degree of interconnection of the ensemble. The remaining part of the stability analysis follows the same line of reasoning as that of Section 9.6.

Detailed analysis shows that for a given N interconnected units, and for fixed values of all chemical parameters, the roots of the characteristic equation are functions of the connectivity number n. As n reaches various critical values, the homogeneous steady state bifurcates into new spatial or spatio-temporal organized states. Thus the connectivity number stands as a new bifurcation parameter. The self-organizing behavior appears not because of chemical changes or volume increase but simply as a result of topological modifications inside the multiply connected system.

These results may be seen more clearly in the following social game. Forty players are given means to earn dollars and francs by some unspecified autocatalytic process. The rule of the game is that each player has to exchange

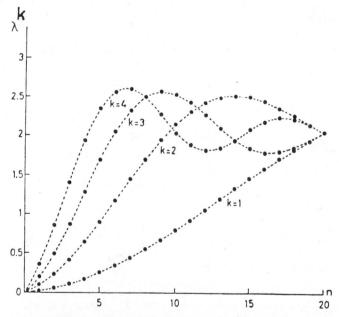

Fig. 14.2. A plot of the first four values of λ^k from Eq. (14.4) against n for a 40-unit ensemble with periodic boundary conditions.

per unit time a fixed amount of these currencies with other members of the society of players in such a manner that the total amount in and out of everybody's pocket remains always the same. For different scenarios however, the number of contacts n of an individual with other members of the society may vary. We assume that, after some time, because of the availability of resources the production and exchange of currencies in the society reaches a "financial" steady state. The outcome is a perfectly egalitarian society. Everyone has the same amount of money. Such a unique steady state exists for all values of n.

Now an individual may receive a few dollars from outside. For some values of n_c the society no longer remains egalitarian; the homogeneous steady state becomes unstable. What happens afterwards is critically dependent on n and also on various fixed economic factors. For each value of n a new self-organizing solution may occur. For example, there are economic cycles of poverty and affluence for all, or a society of very rich and very poor individuals. The important point is that these changes occur in a society where nothing has changed but the number of interpersonal contacts.

Let us go back to a multi-unit and multiply connected chemical system subject to appropriate autocatalytic reactions. The specific kinetics of these reactions inside the isolated units determine the influence of the connectivity number n on the nature of the self-organizing solutions. In general, increasing the connectivity of an ensemble of units with multiple contacts may modify the nature of the spatio-temporal patterns generated by the ensemble. These changes are generally in the direction of a greater stability. Temporal periodicity is nearly always sacrificed by sufficiently increasing the multiplicity of contacts. Under certain conditions, however, purely spatial patterns may arise through an increased number of connections.

*   *

*

In the following sections the properties of multiply connected systems will be used to investigate a network model of the nervous system featuring nonlinear contacts. This model establishes correspondences between spatio-temporal bifurcating solutions of the multiply connected ensembles and various activities of the mammalian brain. But let us first give a short summary of the salient physiological properties of the central nervous system.

14.3. Physiology of the Brain

Science aims at understanding nature and man. But the ultimate in understanding "ourselves" is to encode the modes of functioning of the brain.

This most complex organ is the headquarters of our thoughts, emotions, self-consiousness, and perception. It is the supreme commander of all our actions. A scientific approach to the study of the brain might look like a circular process. Brain research is the only instance in scientific endeavor in which an instrument must investigate itself.

If we discard such philosophical questions as what is "consciousness," "mind," or "understanding" and look at the brain as a multicellular tissue (admittedly of a most sophisticated structure), all the arsenal of biochemistry can be applied, with great success, to its study. The elucidation of all the secrets of the brain is certainly a titanic task; but already we understand some key aspects of brain function in terms of the ordinary laws of physics and chemistry. For example, we may answer such questions as: how are external impulses received and action generated? what is the molecular basis of neural activity? how do nerve pulses propagate in the nervous system?

14.3.1. Brain Structure

The human brain is made of some 10^{11}, interconnected nerve cells called *neurons*. These are arranged in an intricate network of complex structure. Some of the important tasks are performed by cells forming a thin outer layer called the *cerebral cortex* (see Fig. 14.3). No two neurons have the same shape. In the cerebral cortex neurons are organized in space in localized brain functions responsible for seeing, touching, hearing, feeling, hating, loving, and so on. As we shall see, all these activities may be traced to an unromantic propagation of electrical signals generated by the chemical and electrical activity of brain cells.

Every activity of the nervous system is a coherent phenomenon that is generated by the cooperation of a great number of neurons. Here, too, the detailed study of the biochemical activity of a single cell cannot explain the complex and varied behavior of the central nervous system. Therefore, it is tempting to extend the methods of the preceding chapters to neuronal systems. As we shall see model systems may be constructed which account for some aspects of brain function.

14.3.2. Neurons

Brain cells, called neurons, are distinct from other cells of the organism in several respects. They have a characteristic shape, generate electrical current, and communicate simulataneously with thousands of other cells. Another of their peculiarities is that they do not reproduce themselves. On the contrary, we lose about a thousand neurons a day. Apart from these differences, neurons perform all the housekeeping chores of ordinary cells, such as protein synthesis

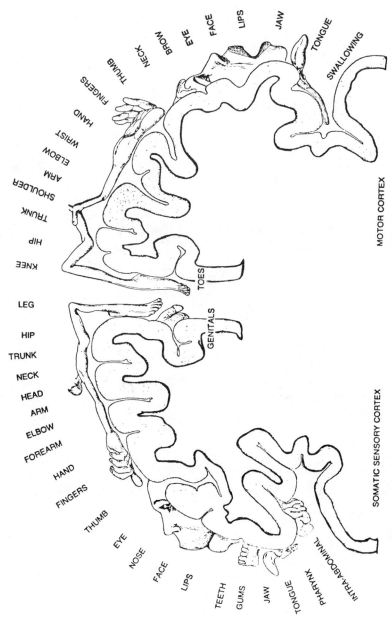

Fig. 14.3. Cerebral cortex. This figure shows the left somatic sensory area (which controls primarily the right side of the body) and the motor cortex (which controls the left half of the body). Most of the body can be mapped onto the cortex. An area dedicated to a specific part of the body is proportional to the precision with which it must be controlled. (From N. Geschwind, 1979, reprinted by permission *Scientific American.*)

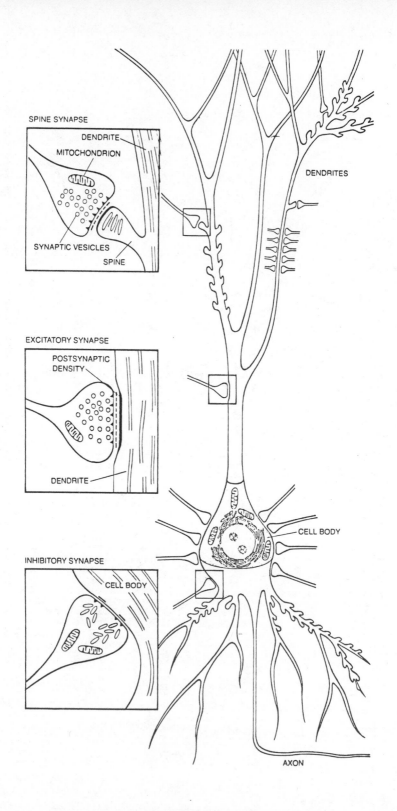

Fig. 14.5.   Cerebral cortex comprises six layers of neurons forming a complex ordered structure. (Courtesy of P. Van Gansen.)

and repair. Neurons are voracious energy consumers. They only burn glucose and use about 20% of the resting oxygen consumption of the organism. The neurons are supported and nourished by a large number of *glial* cells of the brain.

Neurons appear under most unusual shapes. A typical neuron has three distinctive parts, as shown in Fig. 14.4: a *cell body*, an *axon*, and many *dendrites*. The cell body harbors the nucleus and is the locus of all protein synthesis and cell repair processes. The nerve impulse also originates in the cell body. The axon is a long cable-like process, from one millimeter up to one meter long, which stems from the cell body. Axons conduct nerve impulses rapidly and without decrement. The axons branch into a tree-like structure in

Fig. 14.4.   A nerve cell is made of a cell body, an axon, and many dendrites. Synapses may have different positions on the surface of the receiving neuron. Excitatory synapes tend to have round vesicles and a continuous dense thickening of the postsynaptic membrane. Inhibitory synapses tend to have flattened vesicles and a discontinuous postsynaptic density. (From L. L. Iversen, 1979, reprinted by permission from *Scientific American*.)

order to make contact with thousands of other cells. At the opposite end, the cell body extends into many *dendrites*. These are delicate extensions that branch repeatedly up to several thousand filaments and form an intricate network of cells, as shown in Fig. 14.5. In the process of impulse propagation most of the action is in the dendrites.

14.3.3. Cell Electricity

By the late eighteenth century Galvani and Volta were experimenting with electricity detected in the nerves and muscles of the frog. However, the molecular nature of the phenomenon has been elucidated only recently. Today we know that many signaling processes inside the animal body proceed by electrical means. However, the mode of transmission of messages is different from that prevailing in telephone or telegraph wires, since nerve cells are extremely poor conductors of electricity.

The plasma membrane of neurons, like most other cells, is able to maintain within the cell a fluid concentration markedly different from the surrounding liquid. In particular, the interior of the cell has a high concentration of K^+ and a low level of Na^+. These ionic gradients are generated by the Na^+-K^+ pump (see Fig. 12.3) which constantly extrudes sodium ions and exchanges them against outside potassium ions. This nonequilibrium distribution of ionic concentrations across the cell membrane generates an excess of negative ions inside and unbalanced positive ions outside of the cells. In the absence of any nerve activity, a permanent potential difference, called the *resting potential*, of about -60 millivolts is produced between the inside and outside of the low permeable axonal membranes.

Neurons are able to propagate impulses because of the special properties of their axon membranes. The lipid bilayer of a neural membrane is the same as that of ordinary cells. The specificity results from protein channels of K^+ and Na^+ transport which are embedded in the axon membrane (see Fig. 12.3). Input from several thousand cells is conveyed by dendrites to the cell body of a neuron. The latter triggers the lowering of negative voltage difference across the axon membrane which is adjacent to the cell body. This event opens the electrically gated sodium channels immediately ahead of the electrically altered region (direction of propagation). Sodium flows in and triggers an autocatalytic process whereby more sodium channels are opened.

The incoming sodium flow changes in a millisecond the potential inside the axon toward positive values, and we speak of *depolarization* of the membrane. In the next stage the sodium channels close and the potassium channels open; therefore, potassium flows out of the cell. This outflux of K^+ ions brings the

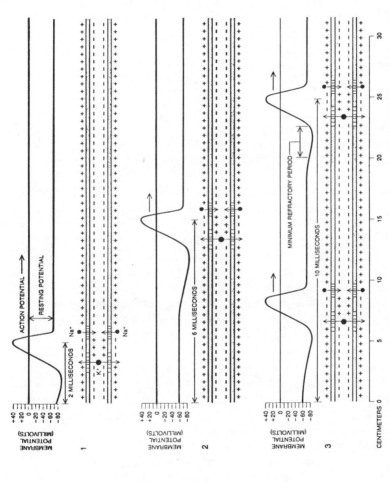

Fig. 14.6. Propagation of nerve impulses. The electrical impulse originates at the cell body. It begins with a small reduction in the negative potential across the membrane which opens sodium channels. The influx of positive ions shifts the voltage still further until a voltage reversal occurs. Sodium channels close and the potassium channels open. The outflow of potassium ions restores the negative potential. An action potential propagates along the axon. Impulses follow each other after a refractory period. (From C. F. Stevens, 1979, reprinted by permission *Scientific American.*)

membrane potential back to a negative value. In two milliseconds the potential reaches $-75\,mV$, which is the equilibrium potential across the membrane in the absence of active transport. In another few milliseconds the K^+-Na^+ pump restores the resting potential value and the membrane is ready for a new impulse.

The sharp positive and negative change of potential is seen as a spike on the oscilloscope, as shown in Fig. (14.6). This kind of nerve activity is called *action potential*. The phenomenon repeats itself all along the nerve axon until the impulse reaches the bulb of the axon terminal and is transferred to the next neuron. The maintenance and restoration of this resting potential requires the expenditure of one third of all ATP of a nonactive organism.

14.3.4. Synapses

Another peculiarity of nerve cells is that information is transferred from one neuron to the next by specific structures called *synapses*. A typical synapse is depicted in Fig. 14.4. The bulb-like axon terminal of one neuron is separated from the dendritic end of another neuron by a narrow cleft. The bulb harbors many round-shaped vesicles, each containing about ten thousand small diffusable molecules called *transmitters*, which are the modulators and regulators of the nervous system. When the nerve impulse arrives from the axon, the vesicles open and release their content into the synaptic cleft. In our example, the transmitter is a small molecule called acetylcholine (see Fig. 14.7). In less than 100 microseconds, transmitters diffuse to the next *postsynaptic* membrane and combine with densely packed acetylcholine receptors, a channel protein which is embedded in the membrane. Two transmitter molecules are needed to change the conformation of the channel proteins from the closed to open state. Sodium ions rush in an potassium ions rush out of these channels, thus *depolarizing* the membrane. This depolarization triggers an action potential in the postsynaptic cell, which in turn propagates the impulse in the manner explained above; the impulse reaches a third cell, and the process continues.

The acetylcholine of the cleft is rapidly destroyed by the enzyme acetylcholinesterase and the postsynaptic membrane returns to its original polarized state and is ready for a new impulse. Meanwhile, the empty vesicles already have been repaired and refilled with newly synthesized transmitters. The number of impulses crossing a given synaptic junction per second is known as the firing rate of the presynaptic neuron. The postsynaptic membrane also becomes polarized, and negative charges accumulate inside the cell.

Synaptic junctions are also made between two dendrites, or between an axon and a cell body. However, there is no propagating impulse in dendrites.

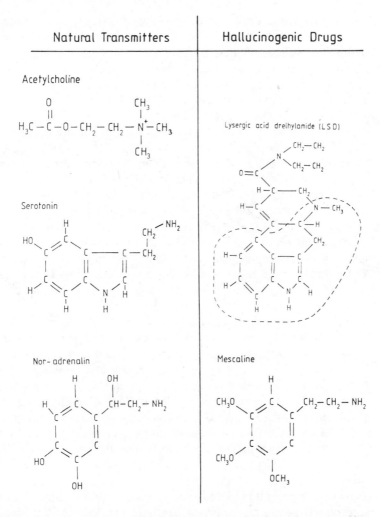

Fig. 14.7. Natural transmitters, and hallucinogenic drugs that show a strong resemblance to the natural molecules. The part of LSD surrounded by broken lines bears a structural resemblance to serotonin.

The incoming impulse to a synapse may be of two kinds. It may promote the firing of the next neuron by making an *excitatory synapse*. In an *inhibitory synapse*, on the other hand, the incoming impulse destroys the effect of excitation of other neurons. The chemical nature of the released transmitter discriminates between these two possibilities. Today we know of some thirty different kinds of transmitters. Figure 14.7 shows a few natural transmitters and also two hallucinogenic drugs.

Membrane receptors are not able to discriminate between natural norepinephrine, dopamine, and the hallucinogenic drugs mescaline and LSD, which explains the harmfulness of these drugs. Daily, we stimulate our synapses artificially by caffeine (coffee) and theophylline (tea). Many conditions, such as nervous tension, depression, or hyperexcitation, may be traced to a dysfunction of synaptic junctions. Most pharmacological drugs as well as toxins temper the normal functioning of synapses. For example, relaxing agents produce persistent depolarization, whereas the psychotomimetic drugs mimic the natural transmitters. The deadly arrows of the Quechuas were covered with curare, which inhibits the depolarization of the postsynaptic membrane and competes with acetylcholine by invading its receptors.

The most honest officer of the antidrug squad lives with a mini-drug plant in his brain. Opiates such as enkephalins and endorphins (from the family of morphine) are synthesized in the brain. These drugs are involved in the perception and integration of pain and emotional experiences, and their receptors also bind morphine. This fact explains the soothing power of the drug. These findings give respectability and a scientific explanation to the Chinese practice of acupuncture, which probably stimulates the synthesis of endogenous drugs.

The events at synaptic junctions are of extremely short duration, of the order of milliseconds. Therefore, in one second several hundred nerve impulses may journey across a synaptic cleft. A given cell body centralizes and performs a spatial and temporal integration of all excitatory and inhibitory inputs (stimuli) received via thousands of synaptic connections. As a result of this integration, the cell body generates a train of impulses with a frequency that reflects the intensity of the integrated stimuli and give rise to nerve impulses in the axon. All impulses have the same amplitude. They are bound by the maximum depolarization potential $-75\,mV$ and the maximum sodium potential $55\,mV$ imposed by the membrane permeability.

The information carried by an axon is characterized by the number of impulses which are generated in unit time. In this *frequency coding* mode of communication, larger the magnitude of the stimuli to be conveyed, the higher the rate of firing. Nerve impulses are proportional to nerve fiber size. At synapses, the fiber size is very small, and the impulse degenerates into local graded potential.

*   *

*

From the standpoint of the self-organizational properties of multiply connected systems, the rudiments of the electrophysiology of the brain reported here may be summarized in the following simple scheme.

The central nervous system of man and animals is made of a collection of individual units with well-defined properties and ionic composition. Synaptic junctions connect units together and form an intricate, multiply connected neural network. The dynamics and the cooperative interactions of the network generate electrical activity in the brain, which is the property of the ensemble of neurons.

In the next section, with the help of such a simple scheme, we model a pathological behavior of the brain and show that the coherent periodic behavior of cells in the course of an epileptic seizure arises as a bifurcating solution from an initially nonorganized state of activity of neurons.

14.4. A NEURAL NETWORK MODEL FOR EPILEPTIC SEIZURES

The first observation of electrical activity recorded from the surface of a living brain was obtained in 1875. Today such recordings are made as a common practice in every hospital for the detection of abnormal structures and functions in the brain. Several pairs of electrodes are opposed between pairs of points on the skull and the brain activity is translated into the motion of a tracer on the paper. In animals, the electrical activity is recorded directly from the surface of the cortex. Such a recording of a normal brain, called an *electroencephalogram, or EEG,* is shown in Fig. 14.8a. Often a pathological state of the brain manifests itself by a modified EEG.

One of the most spectacular brain disorders is the onset of an epileptic seizure. The subject may enter into a trance and lose all consciousness. In the

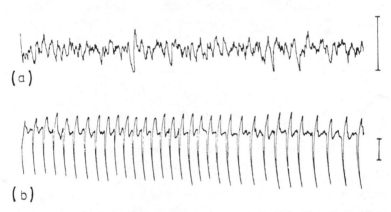

Fig. 14.8. Electroencephalogram (EEG) shows the electrical activity of the brain. Changes in time of voltage difference between two areas of the brain are recorded. (*a*) Normal activity. (*b*) Activity during a seizure. (From L. Kaczmarek and W. R. Adey, 1973, reprinted by permission of Elsevier Biomedical Press B.V.)

Middle Ages the condition was thought to be caused by possession. If recorded, the EEG of the subject shows simultaneous and regular rhythmic firing of a large number of neurons, as shown in Fig. 14.8b.

The exact cause of the disorder is unknown. Sometimes it can be traced to a scar on the surface of the brain. The scar tissue serves as a focus for abnormal activity which spreads into the adjacent normal tissue. Many neurons start to fire with regular rhythmicity. However, natural epilepsy may occur in the absence of scars.

Artificial epileptic seizures may be generated by specific drugs. An example of the EEG recorded during an induced seizure in a cat cortex is seen in Fig. 14.8b. Comparison between the normal and pathological EEG of this figure shows that epileptic seizure is characterized by a large increase in the amplitude of the EEG which manifests extremely regular sharp waves. The firing patterns of nerves also change drastically. Each neuron repetitively gives a high-frequency bursts of action potential followed by a period of inactivation. Many other physiological changes appear in the course of a seizure.

Our present aim is to construct a simple theoretical model in the framework of bifurcation theory which accounts for the two main features, the periodicity and biphasicity of the brain waves during a seizure. At first sight the task seems hopeless, since brain tissue is made of the interconnection of a great number of neurons. However, as we shall see, a reasonable model based on 40 excitatory and an equal number of inhibitory neurons may already account qualitatively for some of the salient features of the experimental observations.

We have seen that the electrical activity of neurons is ultimately based on the physico-chemical properties of various ions and transmitters. however, at the present time a model based on chemical variables seems extremely complex and not yet completely documented. Therefore, we shall construct our model with variables describing the electrical activity of individual neurons. The variables chosen are the mean membrane potentials across the electrically inexcitable dendritic membranes. Let us take $\{x_i\}$ as the potential of N excitatory and $\{y_i\}$ as that of N inhibitory neurons. These membrane potentials are most convenient variables for the description of the system, as the time evolution of the potentials can be directly compared with EEG recordings.

The membrane potential of neurons may change for two reasons. In the absence of synaptic input from other cells, and in the presence of perturbations, the membrane potential of each cell relaxes exponentially to its resting value V with a time constant $1/\kappa$. On the other hand, the excitatory neurons $\{x_i\}$ may receive excitatory input from all N excitatory cells. The effect of the excitatory transmitter is to increase the permeability of the membrane to sodium and potassium ions. During excitatory input, the potential therefore rises toward the transmitter equilibrium potential, denoted by ε. Moreover, the change of membrane permeability of neuron i is proportional to the firing rate of all incoming excitatory cells at the prior time $(t - \tau)$. (τ seconds are needed for an input to reach a second neuron.) The firing rate of the jth excitatory neuron in turn is solely specified by its own mean membrane potential; therefore, it can be written as $f(x_j(t - \tau))$. The explicit form of the firing function f is determined from experimental data and is shown in Fig. 14.9. Thus the dynamics of the network are described by the following equations:

$$\frac{dx_i(t)}{dt} = \kappa(V - x_i(t))$$

$$+ (\varepsilon - x_i(t)) \sum_{j=1}^{N} a_{ij} f(x_j(t - \tau)) + (E_{Cl} - x_i(t)) \sum_{j=1}^{N} b_{ij} f(y_j(t - \tau))$$

$$\frac{dy_i(t)}{dt} = \kappa(V - y_i(t)) + (\varepsilon - y_i(t)) \sum_{j=1}^{N} c_{ij} f(x_j(t - \tau)), \qquad i = 1, \dots, N \quad (14.6)$$

Here the proportionality constants $\{a_{ij}\}$ quantify the strength of inputs to the ith excitatory cells from each of the N excitatory neurons. The third term in the first equation shows the influence of each of N inhibitory cells on the ith excitatory neuron with the coupling constants $\{b_{ij}\}$. The second term on the right-hand side of the second equation represents the excitatory input into the inhibitory cells. Again $\{c_{ij}\}$ represents the strength of the inputs, E_{Cl} is the equilibrium value for chloride ions.

If we follow our usual method of analysis, we must first find the homogeneous steady states of this network. First let us assume that the total input strength of all j excitatory neurons into a given excitatory neuron i is constant. The individual a_{ij}s, however, may differ widely, and a_{ii} may become

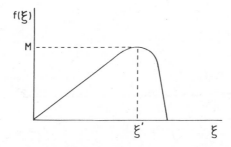

Fig. 14.9. The firing rate as a function of potential is chosen to resemble the curves obtained experimentally. The relationship between firing rate and potential is a linear one up to the level of depolarization. Past this value the relation becomes an inverted parabola with a maximum firing rate M which is obtained when the potential becomes equal to ξ'.

arbitrarily small. The same assumption applies to inhibitory to excitatory or excitatory to inhibitory inputs. Thus:

$$\sum_{j=1}^{N} a_{ij} = A, \qquad \sum_{j=1}^{N} b_{ij} = B, \qquad \sum_{j=1}^{N} c_{ij} = C, \qquad i = 1, \ldots, N \qquad (14.7)$$

Moreover, we assume that at time $t = 0$, all membrane potentials of all excitatory and inhibitory neurons are equal. Subject to these conditions, the $2n$ coupled equations reduce to two coupled sets of N identical pairs which are the same as Eq. (14.6), but with $i = 1$.

The solution of these differential equations is therefore spatially uniform for all $t > 0$. At steady state the two coupled equations are solved numerically for many sets of parameters. A typical result is shown in Fig. 14.10, where the excitatory membrane potential X is plotted as a function of the logarithm of A, which controls the total input of excitatory cells into excitatory neurons. Numerical computation shows that at low levels of excitation there exists only one stationary state. With increasing excitation, multiple steady states appear such that for at least one steady state both x and y take relatively high values. In this case both excitatory and inhibitory membranes are strongly polarized. At still higher values of A only two steady states remain. At the lower state, the

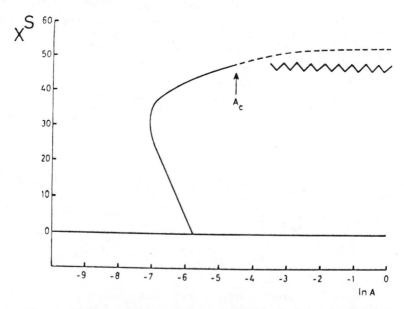

Fig. 14.10. A plot of the homogeneous steady-state value X^s for the excitatory population as a function of $\ln A$. At A_c the upper state becomes unstable (broken lines) and an oscillatory solution appears.

individual neurons do not feel each other's presence and are not polarized. For exactly the same parameters a highly polarized state may also exist. Thus the system is simultaneously offered two totally different possibilities. The upper polarized branch in which neurons influence each other seems more interesting physiologically, and therefore, again following our previous methods, we are led to test the stability of such homogeneous states with respect to arbitrary fluctuations.

The standard linear stability analysis of Section 9.5 must be somewhat modified for the *differential delay* equation (14.6), because of the presence of the time delay τ. One computes straightforwardly a characteristic equation:

$$\lambda^2 + \beta_1\lambda + \beta_2\lambda e^{-\lambda\tau} + \beta_3 + \beta_4 e^{-\lambda\tau} + \beta_5 e^{-2\lambda\tau} = 0. \qquad (14.8)$$

The coefficients $\beta_1, \beta_2, \beta_3, \beta_4$ and β_5 are functions of the steady states and various parameters of the system.

The linear stability analysis shows that, for critical values of $A = A_c$, the upper branch becomes unstable, and therefore the system may bifurcate into an oscillatory solution. Computer integration with $A > A_c$ discloses a limit cycle around an unstable steady state. Close to A_c the limit cycle is of small amplitude, with both types of cells firing continually, $(f(x(t)) > 0, f(y(t)) > 0)$. As A is increased further, away from the critical point A_c, the oscillations take on a different character (see Fig. 14.11). Both types of cells fire in unison, as shown by the black bars above the oscillatory domain marking those regions for which $f(x(t))$ and $f(y(t))$ are different from zero. There follows a period of spike inactivation until the potentials fall sufficiently to start repetitive firing and a new cycle. The oscillations now have also a markedly biphasic appearance, each cycle having a sharp negative-going peak during which the cells are firing, followed very abruptly by a peak in the direction of depolarization.

The transition from rapid continuous firing to the sharp biphasic waves with increasing excitation is directly analogous to the firing patterns of neurons in experimentally induced seizures (see Fig. 14.8). Such oscillatory patterns may also be produced by increasing the values of V and ε, with other parameters held constant. An increase in both V and ε is consistent with a rise in the extracellular concentration of potassium ions in the cerebral cortex.

It is interesting to examine the role of the delay τ in determining the stability of the homogeneous depolarized steady states. The steady-state values of x and y are unchanged by variations in this parameter. For a system in which there is no delay between the firing of a cell and its effect on postsynaptic membranes, the linear stability of a steady state may be calculated by putting $\tau = 0$ in Eq. 14.8. One verifies easily that the system remains stable toward small fluctuations and no oscillations may appear. Clearly, the oscillatory

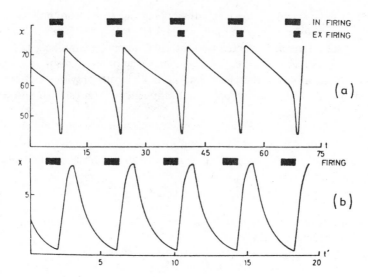

Fig. 14.11. (a) A plot of the sharp biphasic oscillatory solution to the dynamical equations describing the neural network. The times at which the excitatory population fires are marked by the black bars. (b) If inhibitory cells are deleted from the model, oscillatory periodicity remains. However, the biphasic nature of the oscillation is lost.

behavior arises because of the delay τ. For a given set of parameters, with a steady state solution taken on the upper branch it is possible to calculate the critical value of τ_c above which oscillatory solutions may be found.

Equation 14.6 describe a network of $2N$ interconnected cells. Many other types of intercellular connections may be considered in which cells are related to only a fraction of the network population. Again one finds that the homogeneous steady states may become unstable, and various types of spatial, spatio-temporal, and chaotic behavior may appear in the system. Such a chaotic dynamics is shown in Fig. 14.12.

The analogy between the experimental recording of an EEG which shows an epileptic seizure of the cat brain (Fig. 14.8b) and the computed excitatory membrane potentials of a small network (Fig. 14.11a) is striking. The regular time periodicities and biphasic character of oscillations are both recovered by the simple model. Moreover, if the influence of inhibitory cells is discarded from the network by taking $b_{ij} = 0$, $c_{ij} = 0$, then the solutions are purely time-periodic (Fig. 14.11b). This fact suggests that the biphasic behavior of EGG during a seizure is the result of the inhibitory action of neurons, whereas the time periodicity is generated by coherent firing of excitatory cells.

It is encouraging to see that an extremely crude and simplified model, embodying only a few physiological facts, may account for a pathological

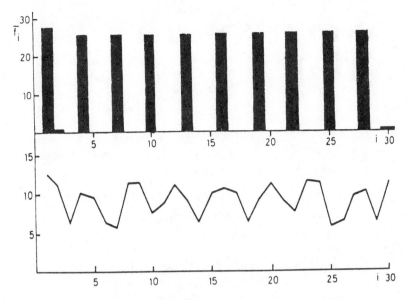

Fig. 14.12. For other ranges of parameters, the neural network model may also generate chaotic behavior.

behavior of the brain function. It is the nonlinear interaction between the units of the network which brings about the coherent behavior of the ensemble. However, the epileptic seizure will not set in unless a threshold value of excitatory input is generated in the network.

*　　*

*

The cerebral cortex is a highly complex network of millions of units with connectivity numbers up to several thousands. Moreover, neurons are organized in spatial and functional units. In light of the methods of the present chapter we may expect a greater number of bifurcation points as well as extremely rich self-organizational properties in a more realistic model of the cerebral cortex. It is also tempting to speculate that most activities of the brain are related to such self-organized states. Thus we feel that the concept of dissipative structures may be used for the investigation of cooperative properties of the central nervous system.

CONCLUSION TO
PART III

The biological examples treated in this section were chosen from among the earliest and simplest instances. However, in recent years the general field of nonlinear phenomena and, more specifically, the investigation of self-organized systems have exploded in many directions. To give even a superficial account of the various domains considered would require the writing of another volume. Such a book, giving a larger view of the various fields of self-organized systems, is in preparation by G. Nicolis and I. Prigogine (*Exploring Complexity*, 1985).

Each year several international conferences in this area gather specialists from different countries. Books and papers are published at an ever—increasing rate. For example, the interested reader may have a fair idea of the diversity of the subjects treated by consulting the 28 volumes of the Springer series in Synergetics. Here we show some general directions of investigation, and also a few references which are relevant to specific problems.

The theoretical study of self-organized systems requires the analytical solution of nonlinear differential equations. A great effort has been made by mathematicians in the construction of analytical solutions of the coupled nonlinear differential equations beyond and around the unstable singular points. A more global and qualitative approach also have been developed. Analytical investigations of some hydrodynamic or chemical systems exhibit, beyond unstable points, many acceptable solutions. However, many of these solutions are not seen in experimental systems. Therefore, one must specify the conditions and the constraints which bring about pattern selection.

Under specific conditions some nonlinear differential equations exhibit chaotic behavior. Such random solutions of deterministic differential equations are presently the focus of much theoretical and experimental research. Chaotic behavior has been observed in fluids and in chemical reactions. The role of chaos is very important in hydrodynamics, as it tends to bring a solution to one of the less understood problems of macroscopic physics, namely, fluid turbulence. The understanding of turbulence is of great importance; the phenomenon is seen in such important fields as climatology, and is a source of waste energy in fluid flow. Chaotic behavior has also been studied in relation to biological systems, such as cardiac rhythms, neural activity, and enzymatic regulation.

The words *fluctuation* or *disturbance* have been used throughout the book, meaning a slight change in the conditions of a system. If such processes are endogenous, random, and spontaneous, we are dealing with fluctuations. The role of fluctuations is of crucial importance for the onset of self-organization in a homogeneous but unstable system. The analysis of fluctuations in nonlinear systems is an important and fast-growing field. Self-organization may arise in some stable systems only if they are perturbed by external fluctuations. The sensitivity of nonlinear chemical reactions to very small external influence is also investigated. Perhaps such a sensitivity is the origin of the choice of L-amino acids in the prebiotic soup.

The rediscovery of the Bray reaction and the exotic properties of the BZ reaction have triggered a search for other self-organizing chemical reactions. Many new reactions have been discovered and their number is increasing constantly.

Hydrodynamic motion can be induced at the fluid-fluid interface as a result of interface chemical reactions as well as of transfer of matter and heat between adjacent bulk phases. Such processes are of importance for the study of cell motility.

The modeling of biological processes as self-organized systems has been extensive. Many different and more elaborate models have been proposed for the types of biological problems we have encountered in the five chapters of Part III. For example, there has been an attempt to account for morphogenesis in embryos by considering mechanical motion. Chaotic dynamics has been shown to exist in brain activity. Multiple steady states have been discovered in several living systems. For example, the dynamics of interaction between a tumor and killer cells of an organism may lead to several distinct steady states. The immune system may also be modeled as a cascade of interacting nodes in a network. One finds several steady states corresponding to the physiological states of the immune system.

The mathematical framework outlined in Part II may be used for the description of many other dynamic systems. This is possible whenever the pertinent variables of the system form nonlinear differential equations. Such problems arise whenever feedback processes are present in the system. Such concepts have been applied with success to such diverse problems as insect societies and urban evolution.

REFERENCES

General

W. Ebeling and R. Feistel (1982). *Physik der Selbstroganisation und Evolution.* Akademie Verlag, Berlin.

P. Glansdorff and I. Prigogine (1971). *Thermodynamics of Structure, Stability, and Fluctuations.* Wiley-Interscience, New York.

H. Haken (1983). *Advanced Synergetics.* Springer, Berlin.

G. Nicolis and I. Prigogine (1977). *Self-Organization in Nonequilibrium Systems.* Wiley-Interscience, New York.

G. Nicolis and I. Prigogine (1985). *Exploring Complexity.* Piper, München.

A. K. Peacocke (1983). *The Physical Chemistry of Biological Organization.* Clarendon Press, Oxford.

I. Prigogine (1962). *Introduction to Nonequilibrium Thermodynamics.* Wiley-Interscience, New York.

I. Prigogine (1980). *From Being to Becoming.* Freeman, San Francisco.

I. Prigogine and I. Stengers (1983). *Order out of Chaos.* Heinemann, London.

CHAPTER 1

C. A. Coulson (1953). *Valence.* Oxford University Press, New York.

I. B. Hart (1923). *Makers of Science.* Oxford University Press, New York.

O. Neugebauer (1962). *The Exact Sciences in Antiquity.* Dover, New York.

F. S. C. Northrop (1984). *Science and First Principles.* Ox Bow Press, New Haven, CT.

R. Rosnick and D. Halliday (1966). *Physics.* Wiley, New York.

N. Uyeda, T. Kobayashi, K. Ishizuka and Y. Fujiyoshi (1978–79). *Chemica Scripta.* **14**, 47.

H. E. White (1966). *Modern College Physics.* (6th ed.). Van Nostrand, New York.

CHAPTER 2

H. E. Avery (1974). *Basic Reaction Kinetics and Mechanisms.* Macmillan, New York.

S. W. Benson (1960). *The Foundations of Chemical Kinetics.* McGraw-Hill, New York.

R. S. Berry, S. A. Rice, and J. Ross (1981). *Physical Chemistry.* Wiley-Interscience, New York.

A. A. Frost and R. G. Pearson (1961). *Kinetics and Mechanisms* (2nd ed.). Wiley, New York.

K. J. Laidler (1965). *Chemical Kinetics* (2nd ed.). McGraw-Hill, New York.

CHAPTER 3

F. C. Andrews (1971). *Thermodynamics: Principles and Applications*. Wiley-Interscience, New York.

H. B. Callen (1960). *Thermodynamics*. Wiley, New York.

K. Denbigh (1971). *The Principles of Chemical Equilibrium*. Cambridge University Press, Cambridge.

E. Fermi (1936). *Thermodynamics*. Dover, New York.

H. C. Van Ness (1983). *Understanding Thermodynamics*. Dover Publications, Inc., New York.

CHAPTER 4

H. B. Callen (1965). In *Nonequilibrium Thermodynamics, Variational Techniques, and Stability*. Chicago University Press, Chicago.

S. De Groot and P. Mazur (1962). *Nonequilibrium Thermodynamics*. North Holland, Amsterdam.

P. Glansdorff and I. Prigogine (1954). *Physica* **20**, 773.

P. Glansdorff and I. Prigogine (1971). *Thermodynamics of Structure, Stability, and Fluctuations*. Wiley-Interscience, New York.

A. Katchalsky and P. F. Curran (1965). *Nonequilibrium Thermodynamics in Biophysics*. Harvard University Press, Cambridge, MA.

J. Meixner (1941). *Ann. Physik* **39**, 333.

L. Onsager (1931). *Phys. Rev.* **37**, 405; **38**, 2265.

I. Prigogine (1947). *Etude Thermodynamique des phenomenes irreversibles*. Dunod, Paris and Desoer, Liège.

I. Prigogine (1962). *Introduction to Nonequilibrium Thermodynamics*. Wiley-Interscience, New York.

CHAPTER 5

P. Glansdorff and I. Prigogine (1971). *Thermodynamics of Structure, Stability, and Fluctuations*. Wiley-Interscience, New York.

I. Prigogine (1962). *Introduction to Nonequilibrium Thermodynamics*. Wiley-Interscience, New York.

CHAPTER 6

G. Ahlers and R. Behringer (1978). *Phys. Rev. Lett.* **40**, 712.

G. Ahlers and R. Walden (1981). *Phys. Rev. Lett.* **44**, 445.

H. Benard (1900). *Rev. Gen. Sci. Pure Appl.* **11**, 1261.

S. Chandrasekhar (1961). *Hydrodynamic and Hydromagnetic Stability*. Clarendon Press, Oxford.

E. L. Koschmieder and S. G. Pallos (1974). *Int. J. Heat Mass. Trans.* **17**, 991.

E. L. Koschmieder (1975). In *Adv. Chem. Phys.* **32**, 109.

H. L. Swinney and J. P. Gollub (Eds.) (1981). *Hydrodynamic Instabilities and the Transition to Turbulence.* Springer, Berlin.

G. I. Taylor (1923). *Phil. Trans. R. Soc. London* **A223**, 289.

M. G. Velarde and C. Normand (1980). *Sci. Am.* **243**, 78.

See also the references cited for Chapter 5.

CHAPTER 7

H. Benard (1900). *Rev. Gen. Sci. Pure Appl.* **11**, 1261.

W. Bray (1921). *J. Am. Chem. Soc.* **43**, 1262.

C. M. Child (1941). *Patterns and Problems of Development.* Chicago University Press, Chicago.

R. Epstein, K. Kustin, P. De Kepper, and M. Orban (1983). *Sci. Am.* **248**, 96.

P. Glansdorff and I. Prigogine (1971). *Thermodynamics of Structure, Stability, and Fluctuations.* Wiley-Interscience, New York.

P. G. Lignola, V. Caprio, A. Insola, and G. Mondini (1980). *Ber. Bunsenges. Phys. Chem.* **84**, 369.

A. J. Lotka (1956). *Elements of Mathematical Biology.* Dover, New York.

D. A. McLulich (1937). *Fluctuations in the Number of Varying Hare.* University of Toronto Press.

H. Poincaré (1881). *J. Math.* **3**, 7.

H. Poincaré (1882). *Les methodes nouvelles de la mécanique céleste.* T.1. Gauthier-Villars, Paris.

A. M. Turing (1952). *Phil. Trans. Roy. Soc. Lond.* **B237**, 37.

V. Volterra (1936). *Leçons sur la théorie mathématique de la lutte pour la vie.* Gauthier-Villars, Paris.

CHAPTER 8

K. I. Agladze and V. I. Krinsky (1982). *Nature* **296**, 425.

B. P. Belousov (1958). *Ref. Radiats. Med. Moscow.*

H. G. Busse (1969). *J. Phys. Chem.* **73**, 750.

B. Chance, E. K. Pye, A. M. Ghosh, and B. Hess (Eds.) (1973). *Biological and Biochemical Oscillators.* Academic Press, New York.

P. de Kepper, A. Pacault, and A. Rossi (1976). *C. R. Acad. Sci. Ser.* t. **282**, C, 199.

R. J. Field (1972). *J. Chem. Educ.* **49**, 308.

R. J. Field and R. M. Noyes (1972). *Nature* **237**, 390.

R. J. Field, E. Köros, and R. M. Noyes (1972). *J. Amer. Chem. Soc.* **94**, 1394; *J. Amer. Chem. Soc.* **94**, 8649.

M. Herschkowitz-Kaufman (1970). *C. R. Acad. Sci. Ser.* C **270**, 1049.

R. M. Noyes and R. J. Field (1974). *Ann. Rev. Phys. Chem.* **25**, 95.

A. Pacault and C. Vidal (1982). *J. Chim. Phys.* **79**, 691.

R. A. Schmitz, K. R. Graziani and J. L. Hudson (1977). *J. Chem. Phys.* **67**, 3040.

J. S. Turner, J. C. Roux, W. D. McCormick, and H. L. Swinney (1981). *Phys. Lett.* **85A**, 9.

J. J. Tyson (1976). *The Belousov-Zhabotinski Reaction*. Springer, Berlin.

B. J. Welsh, J. Gomatam, and A. E. Burgess (1983). *Nature* **304**, 611.

A. T. Winfree (1978). *New Scientist* **20**, 5.

A. T. Winfree (1980). *The Geometry of Biological Time*. Springer, New York.

A. N. Zaikin and A. M. Zhabotinski (1970). *Nature* **225**, 525.

A. M. Zhabotinksi (1974). *Self-oscillating Concentrations*. Nauka, Moscow.

A. M. Zhabotinksi and A. N. Zaikin (1973). *J. Theor. Biol.* **40**, 45.

CHAPTER 9

A. A. Andronov, A. A. Vit, and C. E. Khaikin (1966). *Theory of Oscillators*. Pergamon, Oxford.

J. F. G. Auchmuty and G. Nicolis (1975). *Bull. Math. Biol.* **37**, 323.

——(1976). *Bull. Math. Biol.* **38**, 325.

J. Boissonade and P. De Kepper (1980). *J. Phys. Chem.* **84**, 501.

Th. Erneux and M. Herschkowitz-Kaufman (1975). *Biophys. Chem.* **3**, 345.

——(1977). *J. Chem. Phys.* **66**, 248.

R. J. Field and R. M. Noyes (1974). *J. Chem. Phys.* **60**, 1877.

P. Fife (1972). In *Nonlinear Problems in the Physical Sciences and Biology*. Springer Berlin.

M. Herschkowitz-Kaufman (1975). *Bull. Math. Biol.* **37**, 589.

G. Iooss and D. D. Joseph (1980). *Elementary Stability and Bifurcation Theory*. Springer, Berlin.

R. Lofever and G. Nicolis (1971). *J. Theor. Biol.* **30**, 267.

G. Nicolis and J. F. G. Auchmuty (1974). *Proc. Nat. Acad. Sci.* (U.S.A.) **71**, 2748.

G. Nicolis and J. Portnow (1975). *Chem. Rev. Phys. Chem.* **25**, 95.

G. Nicolis and J. W. Turner (1977). *Ann. New York Acad. Sci.* **316**, 251; *Physica* **89A**, 326.

S. Koga (1982). *Prog. Theor. Phys.* **67**, 606.

Y. Kuramoto (1981). *Physica* **106A**, 128.

N. Minorski (1962). *Nonlinear Oscillations*. Van Nostrand, Princeton, NJ.

I. Prigogine and R. Lefever (1968). *J. Chem. Phys.* **48**, 1695.

J. C. Roux and H. L. Swinney (1981). In *Nonlinear Phenomena*, Springer, Berlin.

D. H. Sattinger (1972). *Topics in Stability and Bifurcation Theory*. Springer, Berlin.

J. J. Tyson and P. C. Fife (1980). *J. Chem. Phys.* **73**, 2226.

A. T. Winfree (1980). *The Geometry of Biological Time*. Springer, New York.

CHAPTER 10

A. Babloyantz (1972). *Biopolymers* **11**, 2349.

M. Calvin (1969). *Chemical Evolution*. Oxford University Press, Oxford.

J. de Rosnay (1966). *Les origines de la vie; de l'atome à la cellule*. Editions du Seuil, Paris.

M. Eigen (1971). *Naturwiss.* **58**, 465.

M. Eigen, W. C. Gardiner, P. Schuster, and R. Winkler-Oswatitsch (1981). *Sci. Am.* **244**, 88.

M. Eigen and P. Schuster (1979). *The Hypercycle: A Principle of Natural Self-Organisation.* Springer, New York.

S. W. Fox (1965). In *The Origin of Prebiological Systems.* Ed. S. W. Fox, Academic Press, New York.

A. Goldbeter and G. Nicolis (1972). *Biophysik* **8**, 212.

J. Hiernaux and A. Babloyantz (1976). *Biosystems* **8**, 51.

J. Keosian (1965). *The Origin of Life.* Chapman and Hall, London.

A. Kornberg (1980). *DNA Replication.* Freeman, San Francisco.

S. L. Miller and L. E. Orgel (1974). *The Origin of Life on Earth.* Prentice-Hall, Englewood Cliffs, NJ.

A. J. Oparin (1965). *The Origin of Life.* Weiden Feld and Nicolson, London.

L. E. Orgel (1973). *The Origin of Life: Molecules and Natural Selection.* Chapman and Hall, London.

I. Prigogine, G. Nicolis, and A. Babloyantz (1972). *Physics Today* **25**, No. 11, 23; No. 12, 38.

E. Schrödinger (1945). *What Is Life?* Cambridge University Press, London.

L. Stryer (1975). *Biochemistry.* Freeman, San Francisco.

P. O. P. Ts'O (1973). *Basic Principles in Nucleic Acids,* vols. 1, 2. Academic Press, New York.

CHAPTER 11

A. Babloyantz and M. Sanglier (1972). *FEBS Lett.* **23**, 364.

J. R. Beckwith and D. Zipser (Eds.) (1970). *The Lactose Operon.* Cold Spring Harbor, New York.

P. D. Boyer (1970). *The Enzymes* (3rd ed.). Academic Press, New York.

A. Fersht (1977). *Enzyme Structure and Mechanism.* Freeman, San Francisco.

A. Goldbeter and R. Lefever (1972). *Biophys. J.* **12**, 1302.

A. Goldbeter and G. Nicolis (1976). *Progr. Theor. Biol.* **4**, 65. Academic, New York.

B. Hess and A. Boiteux (1971). *Ann. Rev. Biochem* **40**, 237.

J. Higgins (1964). *Proc. Nat. Acad. Sci.* (U.S.A) **51**, 989.

F. Jacob and J. Monod (1961). Cold Spring Harbor Symposium. *Quant. Biol.* **26**, 193.

W. T. Keeton (1975). *Biological Science.* Norton, New York.

W. A. Knorre (1968). *Biochem. Biophys. Res. Commun.* **31**, 812.

D. E. Jr. Koshland (1973). *Sci. Am.* **229**, 52.

J. Monod, J. Wyman, and J. P. Changeux (1965). *J. Mol. Biol.* **12**, 88.

A. Novick and M. Weiner (1959). In *Proc. Symp. Mol. Biol.* Chicago University Press.

M. Sanglier and G. Nicolis (1976). *Biophys. Chem.* **4**, 113.

E. E. Selkov (1968). *Europ. J. Biochem.* **4**, 79.

P. Van Gansen (1984). *Abreges Biologie Generale.* Masson, Paris.

CHAPTER 12

J. I. Bonner (1967). *The Cellular Slime Molds.* Princeton University Press, Princeton, NJ.

J. T. Bonner (1974). *On Development.* Harvard University Press, Cambridge, MA.

M. S. Bretscher (1973). *Science* **181**, 622.

G. Gerish (1968). *Curr. Top. Dev. Biol.* **3**, 157.

G. Gerish (1971). *Naturwiss.* **58**, 430.

G. Gerish and B. Hess (1974). *Proc. Nat. Acad. Sci.* (U.S.A.) **71**, 2118.

G. Gerish and V. Wick (1975). *Biochem. Biophys. Res. Commun.* **65**, 364.

A. Goldbeter and L. A. Segel (1977). *Proc. Nat. Acad. Sci.* (U.S.A.) **74**, 1543.

A. Goldbeter and L. A. Segel (1980). *Differentiation* **17**, 127.

W. F. Loomis (1975). *Dictyostelium Discoideum: A Developmental System.* Academic Press, New York.

G. Poste and G. L. Nicolson (Eds.) (1977). *Dynamic Aspects of Cell Surface Organization.* North Holland, Amsterdam.

F. M. Ross and P. C. Newell (1981). *J. Gen. Microb.* **127**, 339.

L. Segel (1984). *Modeling Dynamical Phenomena in Molecular and Cellular Biology.* Cambridge Univ. Press, Cambridge.

S. J. Singer and G. L. Nicolson (1972). *Science* **175**, 720.

F. Stevens (1979). *Sci. Am.* **241**, 48.

M. Sussman (1964). *Growth and Development.* Prentice-Hall, Englewood Cliffs, NJ.

Chapter 13

A. Babloyantz and H. Hiernaux (1975). *Bull. Math. Biol.* **37**, 637.

A. Babloyantz (1977). *J. Theor. Biol.* **68**, 551.

A. Babloyantz and A. Bellemans (1985). *Bull. Math. Biol.,* **47**, 475.

J. Brachet (1974). *Introduction à l'embryologie moléculaire.* Masson, Paris.

V. French, P. J. Bryant, and S. V. Bryant (1976). *Science* **193**, 969.

A. Gierer and H. Meinhardt (1972). *Kybernet.* **12**, 30.

C. J. Grimmilkihyen and H. C. Schaller (1979). *TIBS,* (December), 265

S. A. Kauffman, R. M. Shymko, and E. Trabert (1978). *Science* **199**, 259.

W. T. Keeton (1972). *Biological Science.* Norton, New York.

M. K. Mac Williams (1983). *Develop. Biol.* **96**, 239.

J. P. Murray (1981). *J. Theor. Biol.* **88**, 161.

G. Odell, G. Oster, B. Burnside, and P. Albrech (1980). *J. Math. Biol.* **9**, 291.

A. M. Turing (1952). *Phil. Trans. Roy. Soc. London B.* **237**, 37.

P. Van Gansen (1984). *Abreges Biologie General.* Masson, Paris.

L. Wolpert (1978). *Sci. Am.* **239**, 124.

Chapter 14

A. Babloyantz (1977). *J. Theoor. Biol.* **68**, 551.

A. Babloyantz and L. K. Kaczmarek (1979). *Bull. Math. Biol.* **41**, 193.

T. H. Bullock, R. Orkand, and A. Grinnell (1977). *Introduction to Nervous Systems.* Freeman, San Francisco.

The Diagram Group (1982). *The Brain: A User's Manual*. Berkeley Books, New York.

N. Geschwind (1979). *Sci. Am.* **241**, 158.

L. L. Iversen (1979). *Sci. Am.* **241**, 118.

L. K. Kaczmarek and A. Babloyantz (1977). *Biol. Cybern.* **26**, 199.

L. K. Kaczmarek and W. R. Adey (1973). *Brain Res.* **63**, 331.

B. Katz (1966). *Nerve, Muscle, and Synapse*. McGraw-Hill, New York.

S. S. Kety (1979). *Sci. Am.* **241**, 172.

S. W. Kuttler and J. G. Nicholls (1976). *From Neuron to Brain*. Sinauer.

H. G. Othmer and L. E. Scriven (1971). *J. Theor. Biol.* **32**, 507.

K. Pribram (1983). *Language of the Brain*. Branden House, New York.

F. Stevens (1979). *Sci. Am.* **241**, 48.

P. L. Williams and R. Warwick (Eds.) (1973). *Gray's Anatomy*, 35th ed. Churchill Livingstone, Edinburgh.

Further Readings

P. M. Allen, G. Engelen, and M. Sanglier (1984). In *From Microscopic to Macroscopic Order*. Springer, Berlin.

J. Aschoff (1965). *Circadian Clocks*. North Holland, Amsterdam.

P. Bergé, Y. Pomeau, and Ch. Vidal (1984). *L'ordre dans le chaos*. Hermann, Paris.

G. Carreri (1984). *Order and Disorder in Matter*. Benjamin Cummings. Menlo Park.

E. H. Davidson (1968). *Gene Activity in Early Development*. Academic Press, New York.

O. Decroly and A. Goldbeter (1982). *Proc. Nat. Acad. Sci.* (U.S.A) **79**, 6917.

G. Dewel, P. Borckmans, and D. Walgraef (1984). In *Thermodynamics and Regulation of Biological Processes*. de Gruyter, Berlin.

G. Dewel, P. Borckmans, and D. Walgraef (1984). *J. Phys. Chem.* **88**, 5442.

M. Eigen and R. Winkler (1981). *Laws of the Game*, Knopf, New York.

Faraday Society Symposium (1974). *The Physical Chemistry of Oscillatory Phenomena*. The Faraday Division, Chemical Society, London.

P. C. Fife (1979). *Mathematical Aspects of Reacting and Diffusing Systems*. Springer, Heidelberg.

J. Guckenheimer (1983). *Physica* **70**, 105.

O. Gurel and O. Rössler (Eds.) (1979). *Bifurcation Theory and Applications in Scientific Disciplines*. New York Academy of Science, New York.

H. Haken (1984). *The Science of Structure*. Synergetics Van Nostrand Reinhold.

M. Herschkowitz-Kaufman and T. Erneux (1979). *Ann. N.Y. Acad. Sci.* **316**, 296.

B. Hess and A. Boiteux (1971). *An. R. Biochem.* **40**, 237.

J. Hiernaux and Ch. Delisi (1982). In *Regulation of Immune Response Dynamics*, Vol. 1, CRC Press, Boca Raton.

W. Horsthemke and R. Lefever (1984). *Noise-Induced Transitions*. Springer, Berlin.

D. K. Kondepudi and I. Prigogine (1981). *Physica* **107A**, 1.

R. Lefever and T. Erneux (1984). In *Nonlinear Electrodynamics in Biological Systems*. Plenum, New York.

R. M. May (1974). *Stability and Complexity in Model Ecosystems* (2nd ed.). Princeton University Press.

J. D. Murray (1977). *Lectures on Nonlinear, Differential Equation Models in Biology.* Clarendon Press, Oxford.

A. D. Nazarea, D. Bloch, and A. C. Semrau (1985). *Proc. Natl. Acad. Sci. U.S.* **82**, 5337.

C. Nicolis and G. Nicolis (1984). *Nature* **311**, 529.

G. Nicolis and M. Malek Mansour (1984). *Phys. Rev. A* **29**, 2845.

J. S. Nicolis (1985). *Hierarchical Systems.* Springer, Berlin.

J. Pasteels, J. C. Verhaeghe, and J. L. Deneubourg (1982). In *Biology of Social Insects.* Westview Press, Boulder, Co.

T. Pavlidis (1971). *Biological Oscillators: Their Mathematical Analysis.* Academic Press, New York.

Royal Society London (1981). *Theories of Biological Pattern Formation. Phil. Trans. R. Soc. London.* **B295**.

A. Sanfeld and A. Steinchen (1983). In *Proc. Workshop on Chemical Instabilities at Austin.* Reidel, Dordrecht.

H. Schuster (1984). *Deterministic Chaos.* Physik-Verlag, Weinheim.

L. A. Segel (1984). *Modeling Dynamic Phenomena in Molecular and Cellular Biology.* Cambridge University Press.

K. Shaw (1981). Z. Naturf. **30**, a.

C. V. Sterling and L. E. Scriven (1959). *A.I.Ch.E.J.* **5**, 517.

H. L. Swinney and J. P. Gollub (1978). *Phys. Today* **31**, 41.

C. Vidal and A. Pacault (Eds.) (1981). *Nonlinear Phenomena in Chemical Dynamics.* Springer, Heidelberg.

INDEX